P9-DHI-912

The Great Clowns of Broadway

OTHER BOOKS BY STANLEY GREEN

The World of Musical Comedy
Encyclopaedia of the Musical Theatre
Encyclopaedia of the Musical Film
Broadway Musicals of the 30s (Ring Bells! Sing Songs!)
Starring Fred Astaire
The Rodgers and Hammerstein Story
Rodgers and Hammerstein Fact Book (editor)

The Great Clowns of Broadway

Stanley Green

New York Oxford
OXFORD UNIVERSITY PRESS
1984

Library of Congress Cataloging in Publication Data

Green, Stanley.
The great clowns of Broadway.

Includes index.
1. Comedians—United States—Biography.
2. Theater—United States—History—20th century.
3. Theater—New York (N.Y.) I. Title.
PN2285.G7 1984 792.7'028'0922 [B] 84-9700
ISBN 0-19-503471-6

Printing (last digit): 9 8 7 6 5 4 3 2 1

Printed in the United States of America

Pictures with captions bearing an asterisk
are from the Billy Rose Theatre Collection,
the Astor, Lenox, and Tilden Foundations, the
New York Public Library at Lincoln Center

With love to

Kay, Susan, and Rudy

who make me laugh

Preface

When I first began attending the theatre regularly in the mid-Thirties, I was certain of one basic truth: musicals were the most exciting, glamorous, and musically appealing form of entertainment in the world. They were also the most screamingly funny. For these were the years of our great theatre clowns, the men and women who had come out of circuses and burlesque and vaudeville and night clubs to enliven Broadway with their inexhaustible supply of hilarious antics. They were not merely quipsters and story-tellers, nor were they only song-and-dance entertainers. They were thorough buffoons, totally committed to nothing less than making theatregoers laugh their heads off. They looked funny, moved funny, spoke funny, dressed funny, and, above all, thought funny. They didn't *portray* comic characters, they *were* comic characters. And they were as vital to the very existence of the shows in which they appeared as were the breezy juveniles, the perky heroines, the powerhouse song belters, or the high-stepping chorus.

The ten stage clowns celebrated in this book mostly flourished from the mid-Twenties to the early Forties, though some began their careers as early as the turn of the century and others made appearances as recently as the mid-Sixties. Among comedians discussed were those who eventually succeeded in radio, night clubs, films, and television, but all devoted many years to perfecting their art on the Broadway stage. In fact, durability as well as star quality was a criterion of selection. While the Marx Brothers were unquestionably great clowns, they were primarily identified with vaudeville and movies, with only three Broadway shows *(I'll Say She Is, The Cocoanuts,* and *Animal Crackers)* to their credit. And Eddie Cantor, while he had a lengthy stage career, was more of an entertainer-comedian than a buffoon and did not seem to fit within the limits of this survey.

Though the focus of this book is on Broadway productions, the survey also includes London shows (primarily in the chapter on Beatrice Lillie), as well as those that never made it to New York, touring companies, Ziegfeld cabaret revues (in the Fanny Brice and W. C. Fields chapters), extended-run vaudeville shows (Willie Howard, Ed Wynn, and Victor Moore were in these), and even an ice revue (Joe Cook's last stage assignment). And while most of the productions discussed are musicals, either revues or book shows, performances in nonmusical plays are also covered. Stage credits for all ten comedians featured in the book are found at the end.

Among books and articles consulted: Norman Katkov's *The Fabulous Fanny;* Robert Lewis Taylor's "Comedian" (Bobby Clark) in *The New Yorker;* Alexander Woollcott's "The Legend of Sleepless Hollow" (Joe Cook) in *Cosmopolitan;* Gene Fowler's *Schnozzola* (Jimmy Durante); Maurice Zolotow's *No People Like Show People* (chapter on Durante); Robert Lewis Taylor's *W. C. Fields: His Follies and Fortunes; W. C. Fields by Himself* (edited by Ronald J. Fields); Marc Connelly's *Voices Offstage* (Fields's "Peanut Man" routine); Murray Schumach's "Willie Howard: The World's His Straight Man" in *The New York Times Magazine;* John Lahr's *Notes on a Cowardly Lion* (Bert Lahr); Lewis Funke and John E. Booth's *Actors Talk About Acting* (chapter on Lahr); Gilbert Millstein's "A Comic Discourses on Comedy" (Lahr) in *The New York Times Magazine;* Beatrice Lillie's *Every Other Inch a Lady;* Kyle Crichton's "He Makes Anything Go" (Victor Moore) in *Collier's Magazine;* Keenan Wynn's *Ed Wynn's Son;* Lewis Nichols's "Ed Wynn: Up-and-Coming Actor" in *The New York Times Magazine;* Brooks Atkinson's *East of the Hudson* (chapter on Lillie, Cook, and Wynn).

In preparing this book, I was fortunate to have the opportunity of reading many revue sketches and librettos that were of enormous help in conveying what the comedians did and said on the stage. I am particularly grateful to Don Oliver, who generously allowed me to read scripts from his valuable collection. Reviews in newspapers and magazines were indispensable both for opinions and for descriptions. These I perused with the cooperation of Dorothy Swerdlove and her staff at the Billy Rose Theatre Collection at the Lincoln Center Library, Dr. Mary Henderson and her staff at the Theatre and Music Collection of the Museum of the City of New York, and Louis Rachow of the Walter Hampden Library at The Players.

Others whose knowledge, information, and advice were much appreciated are Gerald Bordman, Louis Botto, Jean Dalrymple, Lee Davis, Benedict Freedman, Ed Jablonski, Robert Kimball, Margaret Knapp, Richard Lewine, Dan Langan, Irv Lichtman, Ian MacBey, Frankie McCormack, Arthur Schwartz, Alfred Simon, Donald Smith, and Kay Swift. I am also eternally grateful to my wife, Kay, for the hours she spent reading the manuscript and for her much-appreciated comments, and to daughter Susan and son Rudy for their interest and moral support. A special word of thanks, too, to my editor, Sheldon Meyer, for his encouragement and guidance, and to Leona Capeless for catching errors and making helpful suggestions.

New York
June 1984

S. G.

Contents

The Great Clowns of Broadway

Fanny Brice

DESPITE PRATFALLS and other physical excesses, despite the frequent requirements of grotesque makeup and bizarre costumes, the orderly disorders of a Broadway clown's world have never been limited to male performers. Within the impressive if limited roster of female clowns two stood out most notably. During the period in which comedy was at least the equal of music in the musical-comedy theatre, no stars were more radiantly comical than Fanny Brice and Beatrice Lillie. Since Miss Lillie was born in Toronto and remained a British subject, it may be fairly claimed that Miss Brice was the premiere funny lady of United States citizenship to be identified with the Broadway stage.

Though she had a somewhat prominent nose (which she later had bobbed), Fanny Brice was a strikingly attractive woman offstage, easily among the best-dressed actresses of her time. She was elegant, poised, and—even with her affinity for salty language—very much a *grand dame* of the theatre. On stage she was prone to the most outlandish outbursts and facial expressions as she distorted her remarkably mobile features into a variety of comic grimaces. With little apparent motivation other than an irresistible urge to cut up, she would cross her eyes, puff out her cheeks, slap her forehead, buckle her knees, and collapse her long, slim body. And even when she wasn't milking a laugh, there was always something ludicrous about the comedienne's glinting, mischievous eyes and wide, half-moon smile that gave her the appearance of a canary-swallowing cat.

For most of her stage career, Fanny adopted a Yiddish accent which she used in revue sketches for comic effect, usually in contrast to the character she was playing rather than for any intrinsic ethnic humor. The humor was simply in the incongruity of hearing a seductive woman—whether an Eve or a Theda Bara or a Pompadour or a Camille—speak with the vocal inflections of the Lower East Side; it was never meant as a putdown of people with accents. Of her other routines that went more directly to her roots, Fanny once wrote, "I never did a Jewish routine that would offend my race because I depended on my race for the laughs. In anything Jewish I ever did, I wasn't standing apart making fun. I *was* the race and what happened to me on stage is what could happen to my people. They identified with me, which made it all right to get a laugh. Because they were laughing at me as much as at themselves."

It was largely through her songs that Fanny revealed her serious side. Almost from her very first appearance on the stage, she was closely associated with specially written numbers that caught both the comedy and the pathos of life in a large metropolis. As she sang, Fanny projected the character of a poor but spunky Jewish kid yearning for romance or fame, or both, often identified by name and, at times, even by the area of the city from which she supposedly comes. Thus she was Sadie Cohen in "Sadie Salome, Go Home," Becky Cohen in "Goodbye, Becky Cohen," and another Becky in "Becky Is Back in the Ballet." Later she was "Soul Saving Sadie from Avenue A" and "Sarah, the Sunshine Girl." The name by which she was best known in song, however, was Rose—as in "Rose of Washington Square," "Rosie Rosenstein (in "I'm an Indian"), "Second Hand Rose (from Second Avenue)," and "Mary Rose." Rodgers and Hart wrote a song for her about "Rest-Room Rose," and Cole Porter gave her "Hot House Rose" (about Rosie Rosenbaum "from God knows where"), though the last piece never entered her repertory. Curiously, Fanny's greatest triumph, "My Man," had begun life as a French song about a year before she introduced the English version in New York.

In her interpretation of most of these songs—as she once said of "Sadie Salome, Go Home"—Fanny put the spirit of "Loscha of the Coney Island popcorn counter and Marta of the cheeses at Brodsky's delicatessen, and the Sadies and the Rachels and the Birdies with the turnover heels at the Second Avenue dance halls."

Often Fanny would combine two styles within one song. When, for example, she sang "You Made Me Love You" in vaudeville, she first delivered the verse and chorus as a heart-tugging emotional ballad. Then, to poke fun at the sentiments she had just expressed, the orchestra would speed up the tempo and Fanny would sing the same number while swinging on the curtain, kicking up her heels, winking and rolling her eyes, and lifting her skirt to show her long legs and knock knees.

Fanny was an imp, a madcap, an uninhibited force whose specialty in revues was doing routines that mocked the artistic pretensions of ballet dancers (pronounced "belly densehs") and the excesses of sirens of the screen or of history or of literature. A hard-driving, self-sufficient woman, she was well aware of the importance of having audiences sympathize with her gaucheries as well as laugh at them. Her attitude toward her craft and her public was once expressed this way:

ROSE OF WASHINGTON SQUARE

Song

Lyric by
Ballard Macdonald
Music by
James F. Hanley

As Introduced by
FANNY BRICE
in the new
Ziegfeld Midnight Frolic
atop the
New Amsterdam Theatre
New York

Price 60 cents

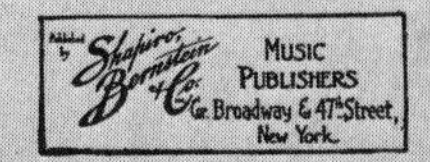

The sheet music cover of one of Fanny Brice's most closely identified songs, which she first sang in 1920.

Fanny Brice and her co-stars, Ted Healy and Phil Baker, in *Billy Rose's Crazy Quilt* (1931).

Willie Howard and Fanny Brice in David Freedman's sketch, "Sailor, Behave!," in the 1934 *Ziegfeld Follies*.

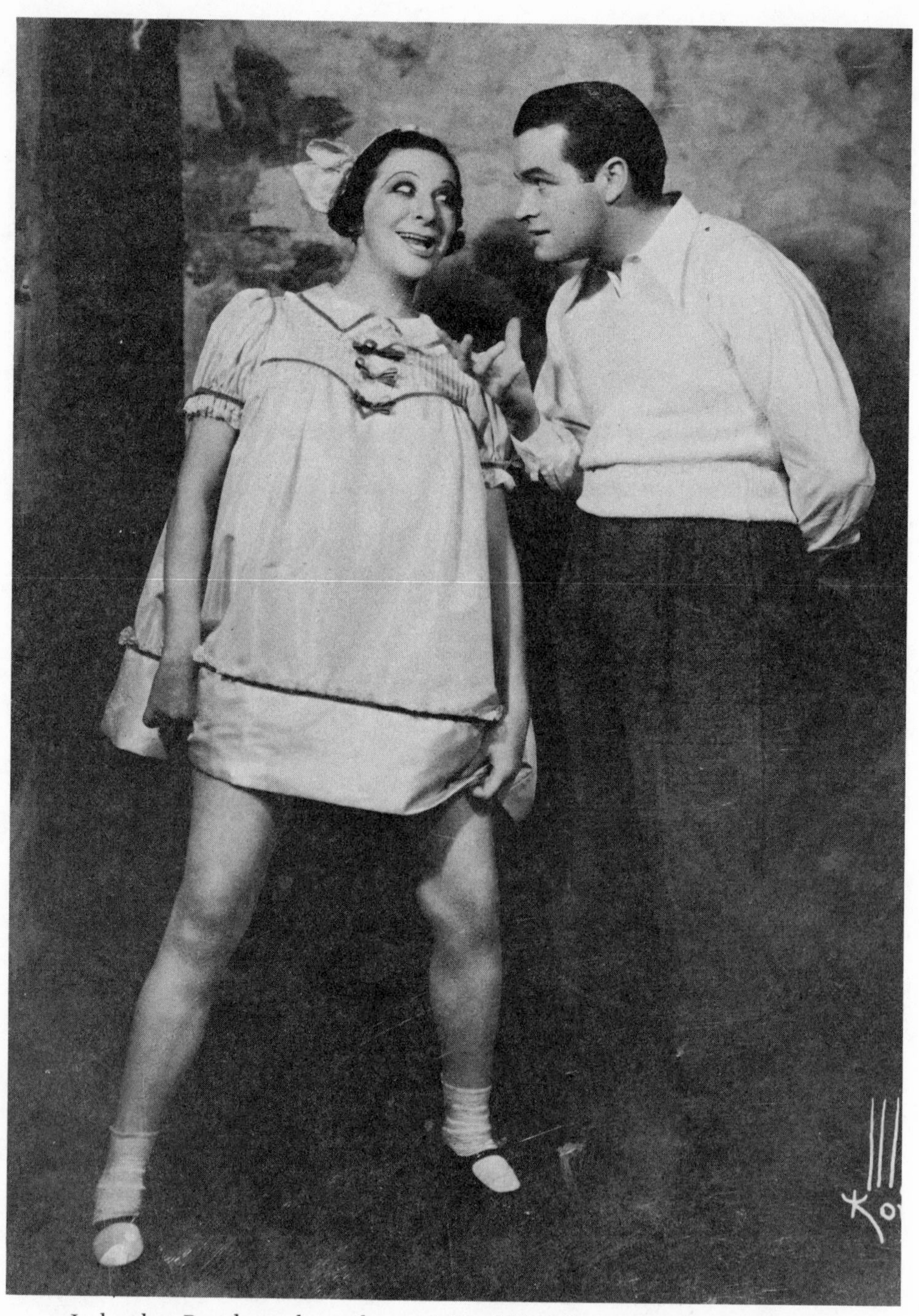

In her last Broadway show, the 1936 *Ziegfeld Follies*, Fanny Brice appeared as Baby Snooks and Bob Hope played a Hollywood director.

You get your first laugh—boom! You're going. You lose yourself, you become whatever it is they're laughing at. But it isn't you. Anytime I ever did any kind of dance, don't think that in my heart, as I'm making them laugh, that I don't want them to say, "She's really so *graceful.*"

Your audience gives you everything you need. They tell you. There is no director who can direct you like an audience. You step out on the stage and you can feel if it is a nervous audience. So you calm them down. I come out before an audience and maybe my house burned down an hour ago, maybe my husband stayed out all night, but I stand there. I'm still. I don't move. I wait for the introduction. Before I do anything, I've got them with me, right there in my hand. And they're comfortable. That's my job, to make them comfortable. Because if they wanted to be nervous they could have stayed home and added up the bills.

If you're a comic you have to be nice. The audience has to like you. You have to have a softness about you, because if you do comedy and you are harsh, there is something offensive about it. Also you must set up your audience for the laugh you are working for. So you go along and everything is fine, like any other act, and then—boom! You give it to them. Like there is a beautiful painting of a woman and you paint a mustache on her.

Fanny Brice (she once said she was billed "Fannie" and signed her checks "Fanny") was born on New York's Lower East Side on October 29, 1891. When she was a child, her parents moved to Newark, where they operated several saloons and where Fanny first got the chance to entertain customers with her songs and dances. Fanny's father, Charles Borach, an immigrant from Alsace, was a bartender with a weakness for liquor and pinochle; her mother, Rose, an immigrant from Hungary, was a strong-willed woman who was the actual head of the family and managed the saloons. It was she who decided to sell the businesses and move the family to Brooklyn, where she did well as a real-estate operator. Fanny first appeared on a real stage at the age of fourteen, when she won $10 in an amateur contest at a Brooklyn theatre singing "When You Know You're Not Forgotten by the Girl You Can't Forget." This was enough to convince her that she was wasting her time in school, and with her mother's approval she quit to become an entertainer. Within a year she was averaging $30 a week by singing in amateur shows around the city.

Before she could be a success, Fanny felt that it was necessary to acquire a new name. From a neighbor, John Brice, she took one that was close enough to the original and far easier to visualize in lights on a theatre marquee. Fanny's first professional engagement, however, was short-lived. In 1917 she was accepted for the chorus of a George M. Cohan musical, *The Talk of New York,* starring Victor Moore, but Cohan fired her because she wouldn't stop cutting up during rehearsals. Fanny's frequently uncontrollable sense of humor, however, was just the thing that did appeal to the manager of the Columbia Burlesque Wheel, the chief burlesque circuit of the day, and he signed her for a part in one of his touring shows. In those years there was no stigma attached to burlesque, which was nothing more than a low-priced, farcical musical show that actually did satirize the fads and foibles of the time.

Needing a song for a benefit show that the burlesque manager wanted her to appear in, Fanny asked songwriter Irving Berlin for something appropriate for the occasion. Berlin invited her to his office, where he played the raggy "Grizzly Bear" and the Jewish dialect number "Sadie Salome, Go Home" ("Don't do that dance, I tell you, Sadie,/That's not a bus'ness for a lady"). "I had no idea of doing a song with a Jewish accent," Fanny later wrote. "I don't even understand Jewish, couldn't speak a word of it. But I thought if that's the way Irving sings it, that's the way I'll sing it. I learned them both in an hour."

For the benefit show, Rose Borach put some extra starch in her daughter's white linen sailor costume. "That sailor suit is killing me," Fanny once recalled. "And it's gathering you know where, and I'm trying to squirm it away, and singing and smiling, and the audience is loving it. They think it's the act I'm doing, so as long as they're laughing I keep it up. They even throw roses at me."

Fanny scored a hit in the Columbia burlesque show, called *The College Girls* ("A Smart, Classy Musical Frivolity in Two Acts"), in which she was described in the program as "Petite Fanny Brice Demure and Charming Soubrette." She did so well, in fact, that by the time the show played New York's Columbia Theatre (then at 47th Street and 7th Avenue) early in 1910, she felt confident that she was ready for Broadway. In that appraisal she was not alone.

At that time, Broadway's most celebrated purveyor of musical entertainment was the colorful impresario Florenz Ziegfeld. Then forty-three, Ziegfeld had been a Broadway showman for fourteen years, with some twelve productions to his credit. Three years before he had begun his series of annual musical revues, the *Follies*, which had won renown for their feminine beauty and opulent decor. Though these shows, as the title indicates, were intended as lighthearted commentaries on current events—social, political, and theatrical—their satire and songs were all too often smothered under the yards and yards of silks and satins that adorned the stately showgirls parading on different levels. Ziegfeld, in fact, cared so little about comedy and music that he delegated these departments to others while he concentrated on the look and "feel" of his extravaganzas. Though the producer himself did not see Fanny at the Columbia Theatre, one of his talent scouts did and strongly urged The Great Glorifier to engage her for the next *Follies*. When she received a telegram from him (Ziegfeld always sent telegrams), Fanny was so sure someone was playing a trick on her that she had to be convinced by her agent that the wire was indeed from Ziegfeld. Not quite nineteen, she signed a contract at $75 a week to appear in the fourth edition of the celebrated Ziegfeld revue.

The Follies of 1910 (the producer's name did not show up as part of the title until the following year) opened in June at the Jardin de Paris located above the New York and Criterion Theatres on the east side of Broadway. Billed as "F. Ziegfeld Jr.'s Song Revue," its cast included the droll Negro comedian Bert Williams (also making his *Follies* debut), voluptuous Lillian Lorraine, singer Rosie Green (Mitzi's mother), and comic Harry Watson. One of the show's highlights was the introduction of motion-picture scenes featuring Anna Held, then Ziegfeld's common-law wife. Fanny sang two dialect numbers, Irving Berlin's Yiddish-flavored "Goodbye, Becky Cohen" and Joe Jordan and Will Marion Cook's "Lovey Joe," supposedly written for Fanny after its creators had enjoyed a sumptuous meal prepared by Fanny's mother.

"Lovey Joe," a not particularly distinguished example of the style once referred to as a "coon song," was treated to such an uninhibited interpretation that Fanny turned it into one of the hits of the show. During rehearsals, however, it almost got her fired. The *Follies* was then a major social as well as theatrical occasion—attracting New York's carriage trade as did no other stage event except a Metropolitan Opera production—and there could be nothing in it that might offend anyone. What did offend A. L. Erlanger, Ziegfeld's chief backer and a successful producer in his own right, was Fanny's Negro dialect, especially the way she slurred the line, "I jes' hollers fo' mo'," and the suggestive manner in which she sang of her affection for "that ever-lovin' man from way down south in Alabam'." During one rehearsal, Erlanger stopped Fanny in the middle of the number and demanded that she sing each word distinctly and with more control as befitting a performer in a $2.50-per-ticket Broadway production. Fanny refused and Erlanger ordered her out of the show. Ziegfeld somehow managed to smooth things over, and Fanny justified his confidence by stopping the show so completely that—legend has it—she was obliged to give twelve encores the first night of the Atlantic City tryout.

From then on Fanny's name was closely linked professionally to that of Ziegfeld and the *Ziegfeld Follies*. She appeared under the showman's sponsorship in seven editions, plus four of his cabaret revues. Even after Ziegfeld's death in 1932, Fanny was featured in two *Follies* produced by the Shubert brothers, in 1934 and in 1936. When M-G-M made *The Great Ziegfeld,* a glamorized life of the impresario, Fanny was the only one of his *Follies* headliners in the movie. And she was the only *Follies* veteran to be featured in the same studio's 1946 release, *Ziegfeld Follies,* repeating the sweepstakes-winner sketch she had first appeared in on stage in the 1936 edition.

Fanny tried duplicating her "Lovey Joe" success in the 1911 *Follies* with her rendition of Irving Berlin's "Ephraham," another southern dialect song celebrating the accomplishments—this time at the piano—of another talented gent in Alabam'. But her appearance did not create quite the sensation that it had in the previous *Follies* and Fanny decided to concentrate on improving her comic technique by touring in vaudeville. It was during this period that she created two spoofing characterizations, the ballet dancer and the vamp (or vampire), that would become something of a trademark throughout her stage career. Fanny soon became such a major two-a-day attraction that she was engaged to play two of the most prestigious variety houses in the world, Hammerstein's Victoria on Times Square and the Victoria Palace in London. While appearing at the Hammerstein theatre, she had a handbill printed showing a demure Fanny Brice in profile underneath the boldface banner heading, "HERE'S SOMEBODY WHO WILL WAKE YOU UP!" Below the photograph was this appraisal: "She's a typical New York girl, as flip and smart as they make 'em in this man's town, and just as foolish as a fox. She's on the corner of 42nd Street and Broadway doing a snappy single."

Late in 1912, Fanny joined the road company of a Shubert production, *The Whirl of Society,* starring the French comedienne Gaby Deslys and an exuberant black-face entertainer named Al Jolson. Fanny's part, which was specially written for her, was that of a Yiddish-accented soubrette. Soon after returning to New York, she played virtually the same role on Broadway in *Honeymoon Express,* another Shubert musical also starring Gaby Deslys and Al Jolson. This appearance gave Fanny four solo numbers, including "My Coca-Cola Belle," inspired by the recently introduced soft drink. Two years later, she took over the leading female role in Jerome Kern's first Princess Theatre musical, *Nobody Home,* just before the show went on the road.

Ziegfeld beckoned again and Fanny joined the cast of the 1916 *Ziegfeld Follies,* which also featured the comic juggling of W. C. Fields, the singing of the beauteous Ina Claire, the humor of the dead-pan black comedian Bert Williams, and the spirited dancing of Ann Pennington. In a scene called "The Blushing Bride," Fanny did a takeoff on the aesthetic pretensions of classical ballet, then introduced a song in praise of Nijinski (who appeared in the unlikely person of the slow-moving Mr. Williams). The sight of Fanny's mock-serious facial expressions, her pigeon-toed, flat-footed leaps, and the contortions of her gangly legs provided some of the biggest laughs of the evening. In another scene, "Puck's Pictorial Palace," leading performers in the revue imitated celebrities of the day as they might be depicted in the weekly supplement to the Hearst Sunday newspapers. W. C. Fields impersonated Secretary of the Navy Josephus Daniels and former President Theodore Roosevelt; Miss Claire did turns as a teary Jane Cowl, an emotional Geraldine Farrar as "Carmen," and a tiny-voiced Billie Burke (Mrs. Ziegfeld); Miss Pennington played innocent Mary Pickford; and Fanny was seen as Theda Bara. At the time Miss Bara was the screen's leading vamp, and Fanny caught all the foolishness and fun of the femme fatale's come-hither eyes, dilating nostrils, heavy breathing, and slinky movement. The next year's edition of the *Follies* marked the debuts of Will Rogers and Eddie Cantor in the series, and it also offered holdovers W. C. Fields, Bert Williams, and Fanny. Here she was a smoldering Egyptian odalisque, and her ballet burlesque combined "Swan Lake" with a Blanche Merrill number, "Becky Is Back in the Ballet" ("Do for mama the dying duck like Pavlova").

Anxious to expand her horizons, Fanny acted in a 1918 "melodramatic farce with songs" called *Why Worry?*. This allowed her to interpolate familiar routines, such as her Theda Bara takeoff, and she also introduced Blanche Merrill's confession of an adopted Jewish squaw, "I'm an Indian" ("Look at me I'm what you call an Indian,/ That's something that I never was before"). Despite Fanny's efforts, the show was not a success. Of the star's performance, Heywood Broun wrote in the *Tribune:*

> The foundation for the play undoubtedly was, and is, the smile of Fanny Brice. It is the longest, loosest, and most fetching grimace of its kind . . . For all its charm there is a shrewdness about it which acts as an ample antidote for any surplus of sweetness . . . Her curious mixture of ungainliness and grace, a sort of ungainliness under perfect control, helps mightily in putting the songs over. She still sings a little better than she acts, although she has a distinct comic gift. It is untrained as yet and she has lapses in which she pounds her points too hard. This is chiefly emphasized by a tendency to make a line a little funnier than it is written by looking crosseyed while saying it.

In 1915, Florenz Ziegfeld had introduced a new theatrical concept—the *Midnight Frolic*, a cabaret revue performed on the top floor of the building on West 42nd Street housing the New Amsterdam Theatre. Since his *Follies* was then being presented in that theatre, the producer frequently tapped headliners from his large-scale revue for his cabaret, where late-night pleasure-seekers could dine, dance, and enjoy a miniature *Follies*. It proved so successful that in 1918 an early show was added called *Ziegfeld Nine O'Clock Revue*. Though she was not then in the *Follies*, Fanny did appear in that *Revue* to sing another custom-tailored piece by Blanche Merrill, "A Yiddisha Vampire." The following year she was in the *Frolic* doing an Apache dance parody and, to the strains of Mendelssohn's "Spring Song," her impression of a Jewish girl's yearnings to be a ballet dancer.

Fanny sang one of her most closely identified numbers, "Rose of Washington Square" (by James Hanley and Ballard MacDonald) in *Ziegfeld Girls of 1920* (a new name for the *Nine O'Clock Revue*) and in the same year's *Ziegfeld Midnight Frolic*. It was something of a trademark song in which Fanny could sing with alternating lump-in-the-throat pathos and raucous humor. First she described—through the eyes of a butterfly no less—a flower so fair that it must blush unseen in the airless confines of a huge city. Then, in the comic version, she turned it into an exuberant, full-throated expression of a madcap Greenwich Village artist's model ("I've got those Broadway vampires lashed to the mast/ I've got no future but oh what a past"). Possibly only Fanny could have made any sense out of it. (When, in 1938, Fox made a film suggesting Fanny's life, with Alice Faye in the lead, it was called *Rose of Washington Square*. Fanny was so disturbed by the similarities that she sued the studio for $750,000. She settled for $30,000.)

The year 1920 also found Fanny again appearing in a *Ziegfeld Follies*. For this edition, she offered the sagas of two more temptresses whom fate has confined to the less than elegant area of Manhattan: "I'm a Vamp from East Broadway" (by Irving Berlin, Bert Kalmar, and Harry Ruby) and "I Was a Florodora Baby" (Harry Carroll–Ballard MacDonald), the lament of the only member of the famous sextet who didn't end up marrying a millionaire ("All the other girls are living fancy,/ My address is 17 Delancey"). And she also did the "I'm an Indian" number first performed in the short-lived *Why Worry?*. In addition to her musical specialties, Fanny was seen in the sketch, "The Family Ford," written by and featuring W. C. Fields. In this saga of a family's outing in a temperamental automobile, Fields and Brice played husband and wife, and Ray Dooley was their bratty four-year-old daughter.*

If there was a watershed production for Fanny Brice, that was unquestionably the 1921 *Follies*, in which her fellow troupers included W. C. Fields, Raymond Hitchcock, Van and Schenck, Florence O'Denishawn, and Ray Dooley. But even in such company Fanny stood out, not alone because of her comic antics but because of "My Man."

"My Man" was not a new song nor had it been written specifically for Fanny. Originally a French jeremiad called "Mon Homme" with music by Maurice Yvain, it had been introduced the previous year in a Paris revue by the celebrated music-hall entertainer, Mistinguett. Initially, Ziegfeld considered having the French diseuse re-introduce the song in the *Follies*, but soon he began thinking increasingly of Fanny Brice. At the time his reasoning had more to do with publicity value than with Fanny's singing ability. Though nothing in her career up to that time suggested that she was equipped to interpret such an unrelieved cry of pain, her private life made it abundantly clear that she had an emotional kinship with the sentiments—now rendered into English by Channing Pollock—that could easily overcome any possible interpretive or vocal hurdles.

In 1913 Fanny had met and instantly fallen in love with a smooth-talking swindler named Nick Arnstein. Even after their marriage six years later, Nick could not keep out of trouble, and early in 1921 he was arrested as the mas-

*For a more detailed account of this sketch, see page 76.

termind of a gang that had stolen five million dollars' worth of securities from various Wall Street firms. Fanny stuck by her man and also borrowed money from mobsters for bail and defense. (Helping Nick was so financially draining that Fanny was rehearsing for the 1921 *Follies* three weeks after her first child was born.)

As soon as she heard the masochistic sentiment of "My Man," Fanny knew that this was one song she had to sing. She found a red wig and put on a black dress, tied a scarf around it, and walked out on the stage for the rehearsal. Ziegfeld threw a fit. He tore the wig off the frightened actress's head and pulled the scarf off the dress. Then he tore holes in the dress, rubbed stage dirt on it, put a shawl over Fanny's head and draped it around her shoulders. Only after he was convinced that Fanny looked appropriately theadbare and miserable did the producer permit her to sing. And she sang it that first time as she would always sing it, simply, directly, without histrionic embellishment, letting the words and the music tell the story that was also her story.

When she appeared in the *Follies* on a bare stage leaning against a lamppost, the theatre's leading funny woman made one of her most indelible impressions revealing the emotions of one whose total devotion had brought her nothing but unhappiness. As Alan Dale wrote in the *New York American*, "Underneath the flickering gleam of a street light, Miss Brice sings of her love for the dock walloper who beats and chokes her and thunders his abuse whenever he comes around. Slushy, mushy hedgetalk. And yet the sheer artistry of Fanny Brice lifts it out of the muck of the commonplace."

The rest of Fanny's material in the 1921 *Follies* was either the expected serio-comic ballad, such as the durable "Second Hand Rose" ("Even Jake Cohen, he's the man I adore/ Has the nerve to tell me he's been married before"), or the all-out slapstick sketch, such as the first-act finale, with Fanny and Ray Dooley offering their farcical version of the impending Jack Dempsey–Georges Carpentier heavyweight championship fight. Here both comediennes could take full advantage of their special gifts for burlesque as they traded blows with exaggerated ferocity and responded to pain with dazed looks, crossed eyes, and buckling knees. The fight ended with everyone on stage—including the referee, the announcer, and the chorus girls seated at ringside—being knocked flat. As she had in the 1920 *Follies*, Fanny again acted in a sketch as W. C. Fields's wife and Ray Dooley's mother. In the scene, which was written by Fields, the members of the Fliverton family experience all manners of frustration in their attempt to board a subway train that will take them to the country for a day's outing.*

Fanny's funniest skit was a spoof of the Barrymores, with Fanny as Ethel,

*For a more detailed account of this sketch, see page 77.

Raymond Hitchcock as Lionel, and W. C. Fields as John. Following a self-parodying song, the triad appeared in the final scene of *The Lady of the Camellias,* one of Ethel's great stage successes. For the properly tearful meeting, our bedridden heroine (Fanny) is entirely covered by camellias as she is visited by her lover (Fields) and his father (Hitchcock). Pale, languishing, and full of remorse, K'meel confesses between coughs, "I've been such a bad woman." Then suddenly sitting up, she peeks through the flowers and adds with unbridled relish, "But I've been awfully good company."

A few days following the *Follies* opening, Fanny's admirer, Heywood Broun, now writing for the *World,* apologized for being insufficiently appreciative of the lady's talents in his initial review. "There was only a brief reference the other day," he wrote, "to the glow of positive genius which Fanny Brice imparts to the interludes in which she appears. Miss Brice is a consummate artist. Each year she achieves her effects by more subtle and delicate means. There is humorous suggestion now to the slightest movement of her hands, in the merest glance of her eyes. Her sense of the comic is unfailing."

For her return to the *Follies* in 1923 (Bert and Betty Wheeler, Ann Pennington, and Paul Whiteman's orchestra were also in it), Fanny sang a similarly staged sequel to "My Man" called "Mary Rose" (also composed by Maurice Yvain). But the actress was so displeased with most of her material that she left Ziegfeld and never again appeared on Broadway under his banner. Late the following year she joined the fourth and final edition of the celebrated *Music Box Revue.* Produced by Sam H. Harris with songs by Irving Berlin, it was presented at the Music Box Theatre (which Harris and Berlin had built specifically as a showcase for Berlin songs) and featured such stellar attractions as the comedy team of Clark and McCullough, soprano Grace Moore, and musical-comedy juvenile Oscar Shaw. Fanny's switch from Ziegfeld to Harris did not result in any switch from her basic material or characterizations. There was a parody on the ballet in the straightforward declaration, "I Want To Be a Ballet Dancer." There was a sketch in which she rolled her eyes and spoofed a seductive Mme. Pompadour, and another in which she played temptress Eve (wearing a snake headdress and outsized splay-toed feet) to Bobby Clark's club-wielding Adam. Her best musical number was the comic plaint, "Don't Send Me Back to Petrograd," in which she appeared in shawl and tattered dress as a forlorn immigrant on Ellis Island who pleads to remain in the United States ("I promise to work the best I can,/ I'll even wash the sheets of the Ku Klux Klan").*

*As soon as the *Music Box Revue* ended its Broadway run in May 1925, Fanny assumed the unaccustomed duties of producer. Because of her fondness for Bert and Betty Wheeler, the comic song-and-dance team with whom she had appeared in the 1923 *Follies,* Fanny co-sponsored a musical comedy, *The Brown Derby,* in which the couple starred. The show, however, closed after its New Haven tryout.

Following a season in vaudeville, Fanny tried breaking out of the revue mold by starring in a play that, through no coincidence, was name *Fanny.* Billed as a "melodramatic comedy," it was written, directed, and produced by the legendary David Belasco. Fanny played her familiar Yiddish-accented vamp—named Fanny Fiebaum—in the alien locale of an Arizona ranch, but she was unable to convince anyone that she was ready to carry a full-length play without the aid of songs and specialties. So after a brief run it was back to vaudeville, followed by the leading role in *My Man,* an early talkie in which she sang old favorites and introduced "I'd Rather Be Blue." But this appearance made little impression on movie audiences, and Fanny returned to Broadway in the unlikely surroundings of an operetta.

Fioretta was a curious and costly footnote in the annals of the Broadway stage. For a wealthy Philadelphia dowager, who put up the entire $350,000 needed to open the show, it was a chance to further the careers of two tyro songwriters. For producer Earl Carroll, who heretofore had made his name as a purveyor of flashy, fleshy revues, it was a chance to make a star of his mistress, showgirl Dorothy Knapp, as well as to invade a more "respectable" area of musical entertainment. And for Fanny Brice, it was another chance to break away from her accustomed revue field and to create a sustained character without resorting to her customary mannerisms.

But the opulent, stately spectacle, set in and around 18th-century Venice, turned out to be a three-hour bore with the further distinction of being the most expensive Broadway flop up to that time. Feverish attempts to enliven the proceedings during the tryout were of little benefit nor was an opening-night performance full of missed cues and malfunctioning scenery. Realizing all was lost, Fanny reverted to her familiar East Side accent, ogling eyes, and broad gestures, and even managed to bring down the house with her comical, if ethnically misplaced, singing of "The Vicked Old Willage of Wenice." Fanny left the show before the end of its three-month run.

Though Fanny had remained loyal to her man as long as possible, she divorced Nick Arnstein in 1927 because of repeated infidelities. Two years later—and just four days after the opening of *Fioretta*—she married songwriter-showman Billy Rose. Rose, who had written some of Fanny's vaudeville material, was anxious to become as renowned a Broadway impresario as Ziegfeld. To achieve this goal he was not above capitalizing on his wife's name or including his own name, no matter how awkwardly, as part of a show's title. His first effort, *Corned Beef and Roses,* teamed Fanny with two vaudeville and Broadway stars, Hal Skelly and George Jessel. Thoroughly panned in Philadelphia, the revue underwent changes in songs, sketches, and personnel (James Barton replaced Skelly), and opened on Broadway in the fall of 1930 as *Sweet and Low.* Though still received coolly, the show ran most of the season chiefly on the name value of the stars and the fact that Rose borrowed heavily to keep it going. Fanny's big number was the Louis Alter–Charlotte Kent "Overnight," an emotional, torchy lament ("Overnight I found you and overnight I lost you") staged in the long-familiar manner of "My Man." But the chief distinction of *Sweet and Low* is that it served to introduce a Fanny Brice specialty that became the comedienne's most closely identified role during the last ten years of her career. Called "Babykins" in the David Freedman sketch, Fanny was seen as a precocious three-and-a-half-year-old in a high chair being examined by her pediatrician (played by Arthur Treacher).

DOCTOR: And what seems to be the matter with you, little girl?
BABY: I'm tired of life.
DOCTOR: Why? Don't you go out and play with other little girls?
(Baby shakes her head sadly.)
DOCTOR: Poor child. With whom do you play?
BABY: Little boys.
DOCTOR: Is there any little boy you're particularly fond of?
BABY: Yes, but he's untrue to me.
DOCTOR: Did he leave you for another little girl?
BABY: No—for a fire engine!

Babykins—whose name would subsequently be changed to Baby Snooks—was a characterization that probably had its origin many years earlier. It is generally assumed that Fanny first got the idea from comedienne Ray Dooley, who had specialized in bratty kid parts both in vaudeville and in revue. Fanny, as we know, had played Ray's mother in two W. C. Fields sketches in the *Ziegfeld Follies* of 1920 and 1921. But Fanny insisted that the concept had originated in 1912 when she was in vaudeville and based it on a child actress in the movies.

Author Freedman, however, claimed that the notion of Fanny playing a baby resulted from an emergency called he received from Billy Rose when *Corned Beef and Roses* was having its disastrous tryout. At the time Fanny had just been fitted with a set of false teeth that made her speak with a lisp. The Freedman-Brice solution: Fanny would play a character that naturally lisped—a decision that led inevitably to the birth of Baby Snooks.

In May 1931, less than a month after closing *Sweet and Low,* Rose unveiled a sequel, *Billy Rose's Crazy Quilt.* Faithful Fanny was in it and her co-stars were now Phil Baker and Ted Healy. Surprisingly, it was Baker's topical monologues that provided the funniest moments in a still less-than-average revue. Fanny did get to sing "I Found a Million Dollar Baby in a Five and Ten Cent Store" (by Harry Warren, Mort Dixon, and Billy Rose) while wearing top hat, white tie, and tails, and her Yiddish-accented, clumsy-footed Peter Pan was also well received.

Fanny was away from Broadway for two and a half years, and when she returned it was in another *Ziegfeld Follies,* though this one was produced after Ziegfeld's death. Because of the late showman's indebtedness, his widow, Billie Burke, allowed the legendary title to be used by the Shubert brothers, her husband's rivals, and even agreed to serve as the nominal producer. Along with Fanny Brice, the cast of the 1934 *Follies* included Willie and Eugene Howard, Jane Froman, Vilma and Buddy Ebsen, Robert Cummings (then known as Brice Hutchins), Everett Marshall, Eve Arden (in her Broadway debut), and ballerina Patricia Bowman. After a succession of Broadway flops that had left many wondering if she would ever again find a show worthy of her talents, Fanny was welcomed by some of the most enthusiastic notices of her career. "Better than ever," "in finest form," "at her funniest"—and variations thereof—were the expressions found in the notices written by the rapturous first-night critics.

Though Vernon Duke and E. Y. Harburg wrote most of the songs in the show, all of Fanny's musical specialties were the contributions of Joseph Meyer, Ballard MacDonald, and husband Billy Rose—and all were new approaches to well-tested Fanny Brice routines. In a takeoff on evangelist Aimee Semple MacPherson, she was a Yiddish-accented "reformed substitute" known as "Soul-Saving Sadie from Avenue A" ("Vending salvation and making it pay"). As "Sarah, the Sunshine Girl," she revealed the uninhibited joys of being a "noodist." And as "Countess Dubinsky," without her kilinski, who's showing her skinski in a burlesque by Minsky, she was up to her old shtick of satirizing the femme fatale as she performed both a strip tease and a fan dance. Trying to manipulate her two huge fans as gracefully and seductively as possible while playfully striving to epitomize lascivious allure, Fanny managed to get her movements frantically—and hilariously—mixed up. "Here is burlesque with the bite of satire in it," wrote Robert Garland in the *World-Telegram.* "Here is good mean fun. Every little movement has a meanness all its own. You must look sharp to catch on to some of them. The lift of a knowing eyebrow. The lift of a protesting knee. The lift of a feather fan. These and fan dancing is as good as done for."

The revue also marked the first time that Fanny Brice acted in a sketch in which she used the name Baby Snooks. In the scene, also written by Freedman, the child's long-suffering parents (played by Eve Arden and Victor Morley) try vainly to get their offspring to tell the truth by relating the story of George Washington and the cherry tree. Then the child's father turns to the more specific problem of a broken window.

FATHER: Now tell me what happened.

SNOOKS: Well . . . I take out a bottle of ink . . . to write a letter . . . *(sudden inspiration)* and a big lion jumped in through the window and scared me.

FATHER: What!! A lion jumped in through the window?

SNOOKS: Yeah . . . and that's how the window got broke.

FATHER: Now Snooks, even if a lion did jump in through the window, how is it that the pieces of glass are all on the outside?

SNOOKS: Well, the lion jumped in backwards.

FATHER *(furious):* Go on! What happened after the lion jumped in?

SNOOKS: So I ran to the other corner of the room . . . and there were four more lions . . . and they all jumped on me, daddy.

FATHER: And then what happened?

SNOOKS: I got killed!*

The 1934 *Ziegfeld Follies* did well enough for the Shuberts to again secure the rights to produce a new edition in 1936. Once more, there was an impressive lineup of talent, with Fanny leading a company that included Bob Hope, Josephine Baker, Gertrude Niesen, Harriet Hoctor, Eve Arden, Judy Canova, Cherry and June Preisser, and the Nicholas Brothers. The show was even more enthusiastically received than the previous edition. So was Fanny. Critic John Mason Brown wrote, "Although one gets used to writing such words as these about her, I doubt if Miss Brice has ever been funnier. She is given many chances to bring her famous slice-of-honeydew-melon smile, her delicious mimicry, her occasionally crossed eyes, her flat-footed capers, and her knees that often are not on speaking terms with one another to skits and songs which gain enormously because of her ever-hilarious presence."

By 1936 Fanny had separated from husband Billy Rose, and her specialty numbers in the *Follies* were written by the chief creators of the score, Vernon Duke and Ira Gershwin (their "I Can't Get Started," soon to become a standard, was sung in the show by Bob Hope to Eve Arden). In "He Hasn't a Thing Except Me" ("I give you his highness, a pain worse than sinus"), Duke and Gershwin presented Fanny with her first chance to do an out-and-out spoof of her "My Man"-type torch song. She began the threnody leaning against a lamppost, but once she began to sing even the lamppost withdrew its support and walked off the stage. Later, in the middle of the song, she stopped singing to explain, "Well, you get the idea. You know, I've been singing about this bum for fifteen years under different titles. He's always the same lowlife, always doing me dirt but I keep loving him just the same. Can you imagine if

*See page 87 for Willie Howard's story in *The Show of Wonders* of *his* encounter with a lion.

I really ever met a guy like that, what I would do to him! Why I'd . . . It's no use talking . . . That's my type."

Replacing Fanny's ballet burlesques was an avant-garde variation, performed to the song "Modernistic Moe," in which Fanny lamented having married a radical whom she could please only by performing impressionistic modern dances (punctuated by her battle cry, "Rewolt! Rewolt!"). The sketches, all by David Freedman, were among the strongest she ever had. Baby Snooks was back, this time as a tough little screen rival to Shirley Temple who infuriates director Bob Hope by being unable to remember even the simplest lines in her script. To please her the studio even secures Clark Gable and Joan Crawford to play her parents, with Greta Garbo as her nurse. Another sketch, "The Sweepstakes Ticket," cast Fanny as the holder of the winning Irish Sweepstakes number whose husband (Hugh O'Connell) unknowingly gives away the ticket to their landlord (John Hoysradt) as part of their rent money. Such a situation gave Fanny full opportunity to display a wide range of frantic emotions—at first disbelief and uncontrollable joy at her good fortune, then tongue-tied grief when she discovers what her husband has done, followed by shameless flattering, vamping and cajoling the landlord in a frenzied attempt to get the ticket back.*

The David Freedman–Ira Gershwin first-act finale, "The Gazooka" (Hollywood's latest dance sensation), kidded the recent crop of stereotyped backstage movies, with Fanny as Ruby Blondell, Bob Hope as Bing Powell, and Gertrude Niesen as Dolores Del Morgan starring in the latest extravaganza, *The Broadway Gold Melody Diggers of 42nd Street.* A highlight of the sketch was the unreeling of coming attractions for this "Technique-Color Widescope Screen" release, with innumerable credits as well as tantalizing slogans flashed on the screen:

MORE LAUGHS THAN "ANNA KARENINA"

MORE SONG HITS THAN "DAVID COPPERFIELD"

MORE DRAMA THAN "HORSE FEATHERS"

In the encapsulated yarn, Bing is a naïve young songwriter who leaves his small town to make his way in New York. There he quickly meets and falls in love with Ruby, showing his love by putting $112 into a show provided that Ruby is hired for the chorus. The producer (Hugh O'Connell) agrees without

*A close facsimile of the stage performance may be seen in the 1946 M-G-M film *Ziegfeld Follies*, when Fanny repeated her performance with Hume Cronyn as her husband and William Frawley as the landlord.

hesitation ("Rehearsals are in five minutes. We open tonight."). Bing becomes temporarily infatuated with the show's star, Dolores Del Morgan, who drops her hankie for him to retrieve ("Thank you. 271 Park Avenue. Penthouse. Cocktails."), but he returns to heartbroken Ruby after she pleads, "Are you going to let silks and satins rob you of your heavenly gift to knock off hits?" Just before the opening-night performance, however, a new crisis develops.

PRODUCER: Good God! No show tonight.

BING: Why not?

PRODUCER: The leading lady can't find her teeth.

BING: Ruby has teeth. She can play the part.

PRODUCER: OK.

RUBY: I'll play it on one condition. You must let me sing the new song that Bing wrote.

PRODUCER: What is it?

RUBY: It's another "Carioca." It's another "Continental." It's called "The Gazooka."

PRODUCER: Let's hear it.

RUBY *(singing):* "What's the latest rage/ On radio, screen or stage?"

PRODUCER: Stop! It's great! It's another "Blue Danube." This song will make the show a big hit. I think we can all get married.

In another well-received skit, "Fancy Fancy," Fanny and Bob Hope were stiff-upper-lip Britons, Zuleika and Sir Robert, confronting an emotional crisis precipitated when the titled gentleman proposes marriage.

ZULEIKA: Oh, no, Sir Robert. Anything but marriage. It'll be so much trouble divorcing Sir Henry. I'll be your mistress, though. We could have frightful fun and all thet and whet net.

SIR ROBERT: But demmit, girl. I already have a mistress. What I want's a wife. Opposite ends of the dinner table, what ho, and all thet.

ZULEIKA: How too tiresomely decent of you, old platter of kippers, but I couldn't think of it.

The two then break into a song-and-dance number in which the lady confesses her adoration for such heroes as Jack Buchanan, Fred Astaire, Max Baer, Ramsay MacDonald, Cecil Beaton, Noël Coward, and Leslie Howard and ends by admitting:

If you were like the fancies I fawncy
I fawncy I could fency fency you.

At the time, Fanny was suffering from spinal neuritis, which forced the 1936 *Follies* to close in May. Once Fanny's health improved, the show reopened in September with a somewhat different cast: Bobby Clark replaced both Bob Hope and Hugh O'Connell, Jane Pickens took over Gertrude Niesen's songs, and Gypsy Rose Lee filled in for Eve Arden. Brice and Clark were judged an especially strong combination, but it was still Fanny's show. As Gilbert W. Gabriel wrote in the *American:* "She is even more elastic of jaw than she used to be, more evilly funny about her slithering eye-glances, and I continue to nominate her the most uproarious clown that ever came out of baby clothes—and then went right back into them."

The *Ziegfeld Follies of 1936* closed on January 19, 1937. It marked Fanny's final appearance on Broadway. Now increasingly pained by illness and depressed over the termination of her marriage to Billy Rose, she decided that the only possible cure would be a change of atmosphere and climate. Settling into the less frantic life of Los Angeles, she appeared in supporting roles in a few films but devoted most of her time to starring in her own radio series that brought Baby Snooks into American homes for more than ten years. Hanley Stafford played her long-suffering father in all the episodes.

Fanny Brice died May 29, 1951, of a cerebral hemorrhage, five months short of her sixtieth birthday. Though she enjoyed security and fame from her broadcasts—and would win even greater celebrity through Barbra Streisand's depiction of her life in *Funny Girl,* both on stage and screen—the real Fanny Brice, like all great comedians, belonged to the theatre. She knew how important it was for her to be near her audience, and she also knew that it was a mutual need. "I made a deal with the audience every time I came out," she once said. "I look at them, I smile at them, and I tell them—and by looking at me they know—that this is really a private party. It's just between them and me."

Bobby Clark

LIKE GROUCHO MARX, he chomped on cigars and moved in a crouching position. Like Harpo, he carried a cane and chased girls. Like Bert Lahr, he emitted guttural growls. Like Joe Cook, he had a beaming ear-to-ear smile. Like Jimmy Durante, he thrived on comic hysteria. Like Ed Wynn, he was addicted to props and outlandish costumes.

But Bobby Clark was no synthesis of other clowns. From his painted-on eyeglasses to his stubby body and bandy legs he was nothing less than the embodiment of the Broadway clown. Bobby's eminence, in fact, rested solely on his stage appearances. Other buffoons may have won fame in nightclubs, radio, films, or television, but Bobby was master of one field only. As early as the mid-1920s he was hailed by critics as a superior low comedian, and by the 1940s they had lost all restraint, with such benedictions as "one of the great blessings of the theatre" (Louis Kronenberger), "one of the funniest men in the history of mankind" (Richard Watts, Jr.), "the funniest man of his age and weight in show business" (John Chapman), and "in versatility, happiness and vitality he is the greatest of the theatre's buffoons" (Brooks Atkinson).

Bobby's career as a Broadway performer spanned more than twenty-seven years, from the *Music Box Revue of 1922* to *As the Girls Go*, which closed early in 1950. Two years after his last New York appearance he was directing the sketches in *Michael Todd's Peep Show*, and six years after that, when he was almost seventy, he was playing the role of Mr. Applegate in the touring company of *Damn Yankees*.

For the first thirty-one years of his professional life, Bobby was teamed with his burly, giggling, hand-wringing stooge, Paul McCullough, whose ratty raccoon coat, college pennant, and wispy mustache were almost as well-known comic trademarks as Bobby's glasses, cigar, and sawed-off cane. Though they shared equal billing, McCullough was always the "feeder," the one who fed Clark the lines or the business that set up his laugh-provoking response.

McCullough was considered by many to be an expert at this self-effacing task and, during their years together, he was generally commended for his role in heightening his partner's effectiveness. He did not, however, prove to be indispensable. After McCullough's death in 1936, by suicide, Clark went on to his greatest period of critical and popular acclaim.

Bobby Clark never walked. He scampered, skittered, and scurried, and sometimes he loped. But he never walked. No one else ever seemed so hell-bent on dashing about furiously without any apparent destination. Often he would start off madly in one direction, check his course, career with equal determination in another direction, and end up just about where he started. A short, squat man, no taller than 5′4½″, he gave the impression of even shorter stature because of the crouching posture he assumed while filling the stage with frenetic, meaningless, and hilarious movement.

Though he delighted in wearing outlandish costumes, Bobby Clark did have certain props with which he was closely identified and which he used in almost every appearance. Foremost were the greasepaint glasses that he had begun painting on his face during his early years as a circus performer (the reason was either because he couldn't find his regular tortoise-shell glasses or because painted glasses presented no problem for an acrobat). Bobby was famous for his entrance which, with variations, was repeated throughout most of his Broadway career. Out he would come scampering down to the footlights wearing a short covert-cloth top coat and pork-pie hat, with a cigar in one hand and a cane in the other. Beaming from ear to ear, the glint of mischief in his bright blue eyes, he would take a few puffs on the cigar, spit it out, retrieve it, and repeat the business a few more times. When McCullough was with him, he would drop the cigar and wait for his partner to make a grab for it. The minute that happened, Bobby would fend him off with an adroit slashing movement of his cane, then manipulate the cane to get the stogie back into his mouth. He continued to use the routine after McCullough's death, though the prized cigar (which was always greased for easy spitting) was now usually rescued from the clutches of a female member of the cast. If the girl was pretty and scantily clad, he might also signal his carnal desires with a cackling laugh or by pawing the ground and letting loose with a gargling, guttural sound.

In some respects Bobby's comic personality was closest to that of Groucho Marx. There was the low-slung gait, the use of a cigar to punctuate gags, the grease-painted facial adornments, the lascivious look at the girls. Often both men appeared as fast-talking, somewhat shady characters whose impudence somehow managed to be accepted in the highest strata of society or political power. Though their behavior was equally outrageous, the two men approached their characterizations from opposite directions. Bobby was always the hayseed rogue with a face that shone with beaming good nature, while Groucho was the sidewalk sharpie whose face revealed the insincere schemer. Whatever hurdles had to be overcome, Bobby looked like he was having the time of his life. Groucho, however, was usually irritated, disdainful, and, well, grouchy, with a smile used only when he wanted to mask his true knavery. Bobby's leer was innocent and playful; Groucho's leer was more of a sneer. Bobby, it should also be noted, won his greatest acclaim when starred alone, whereas Groucho always had the comic contrast of his more sympathetic brothers, Harpo and Chico.

Other clowns, among them Ed Wynn, Bert Lahr, and Victor Moore, were able to make the transition from buffoon to serious actor. Bobby Clark was the only one to jump from modern buffoon to classical buffoon, with highly praised performances in Congreve's *Love for Love,* Sheridan's *The Rivals,* and Molière's *The Would-Be Gentleman.* Even the purists acclaimed Bobby's ability to be faithful to the source while still projecting his own well-established personality.

Despite his seemingly spontaneous joy on the stage, Bobby—like most great funnymen—was a worrier who fretted for days over the smallest details of a routine. To him the comic business rather than the comic line was all important, and he drew upon his early circus training to help perfect his uncanny feats of tumbling, prop tossing, and pantomime that proved so essential in his brand of knockabout comedy. Irreverent, bawdy, unintimidated, and uninhibited, he came across as naïve and harmless no matter how often he applied his cane to the rear of a retreating showgirl or deceived his pursuers with a wildly ridiculous disguise.

Robert Edwin Clark was born in Springfield, Ohio, on June 16, 1888. His birthplace was the rectory of the First Episcopal Church, where his paternal grandfather served as sexton. Bobby's father, a dour railroad conductor, died when the boy was six. His early fascination with bizarre costumes began when he saw his grandfather, an inveterate collector of Masonic degrees, outfitted in his robes and fraternal insignia. In his youth, Bobby sang in the church choir and learned how to play the bugle. In the fourth grade he became friends with Paul McCullough, four years older, and together they worked up a bugling and tumbling act which they performed at local functions. Their receptions were so encouraging that at the ages of seventeen and twenty-one respectively they decided to try their luck as a professional team. As a result of advertisements they placed in theatrical trade papers, the young men were hired by a minstrel company at a salary of $25 a week for the pair. According to their contract, their assignment was to blow their bugles during minstrel parades, perform an acrobatic act during intermission, sing a few songs, and

Bobby Clark and Fanny Brice as Adam and Eve in the *Music Box of 1924*.*

* *See copyright page.*

Paul McCullough and Bobby Clark in *Strike Up the Band* (1930).*

The sheet music cover (designed by Peter Arno) of Clark and McCullough's 1931 musical, *Here Goes the Bride*.

The Playbill cover of *The Rivals*, in which Bobby Clark co-starred with Mary Boland and Walter Hampden (1942). (Playbill ® is a registered trademark of Playbill Incorporated, N.Y.C. Used by permission.)

Michael Todd's 1944 production, *Mexican Hayride*, found Bobby Clark hailed as the "Amigo Americano."

Robert Shackleton, June Knight, and Bobby Clark in a scene from the 1947 revival of Victor Herbert's operetta, *Sweethearts*.*

do whatever acting was required. They were with the troupe for twelve weeks, received no pay, and were left stranded in Harrington, Delaware. A second minstrel engagement also left them penniless and stranded.

The team's first real opportunity came when a circus took them on as acrobats and clowns. It was from this experience that Clark first concluded that his future lay in comedy. "I think hearing the laughter from that big, dark circus crowd had something to do with it," he once said. "Long before I'd even thought about it, I guess I was feeling that the most important thing in the world was to make people laugh." Initially the laughter came from audiences watching McCullough take spills during the acrobatic act, but Clark soon began adding comic touches of his own. Since he also had to think up things to do as a clown, he began improvising routines and devising odd costumes and makeup, including the painted-on glasses.

One of the team's earliest routines had McCullough climbing onto a table with Clark attempting to hand him a chair. Somehow, the chair would get stuck on the edge of the table, and the rest of the act would involve their sweatful efforts to get it unstuck. By the end of the skit, the boys had taken off almost all their clothes, the table had collapsed, and they had gotten into a fistfight. Another specialty was Clark's imitation of little-known animals. "An antediluvian oyster from the south of Bolivia," he would announce. "I call your attention to the rigidity of its muscles." The "imitation" that followed was simply Bobby falling down in a limp heap. He then would top this by bellowing, "A wild antediluvian Southern-Bolivian oyster calling to its mother!"—and proceed to emit a series of the weirdest cries he could make. By this time, too, Bobby had added a cigar to his props, which he'd move about in his mouth to give comical emphasis to his words.

By 1906 Clark and McCullough were members of Ringling Brothers Circus, and withal they were with various tanbark troupes for about six years. Because circus acts were largely "dumb" acts, requiring no spoken dialogue or jokes, Clark became convinced that they were the quickest route to theatrical oblivion. The partners' future, clearly, had to be in vaudeville or burlesque, where they could make use of greater verbal humor. Clark and McCullough's first vaudeville appearance was in December 1912, in New Brunswick, New Jersey. For their new career they got themselves dolled up in costumes that proclaimed them as members of the shabby gentility: Clark appeared wearing a battered top hat, skin-tight pants, a faded finger-tip length top coat, and a pair of scruffy felt boots, and McCullough was in a seedy fur coat, a straw hat, a tiny string tie, a checkered suit, and a pair of white shoes. The act consisted mostly of well-tried routines they had performed in the circus, including their dependable chair-and-table specialty.

The boys had steady work in vaudeville, but they were never headliners. During this period they began doing fewer acrobatic numbers, relying increasingly on sight gags with dialogue written mostly by Clark. Their biggest success was the lion-tamer sketch in which Clark, as a circus roustabout, volunteers to enter an empty cage if McCullough agrees to don a lion's costume and play the part of the fearsome beast. Unnoticed by Bobby, a real lion enters the cage, and the comedian, while still outside, begins whacking it with his cane through the bars. Once he enters the cage and sees McCullough in his lion costume outside, Clark becomes a hilarious blur of frenzied fear as he rushes around trying to keep out of the animal's reach.

After five years of touring in vaudeville, Clark and McCullough were signed as top bananas of the Columbia Burlesque Wheel. They soon discovered that the madcap, uninhibited world of burlesque, where the emphasis was then on slapstick satire rather than on nudity, was ideal for their brand of comic mayhem. One of their most popular new routines, written by Clark, gave Bobby a chance to show his acrobatic skills in a prizefight sketch. As a timid boxer pitted against a snarling pug twice his size, he used a trampolin concealed in the center of the ring to spring up above his adversary's head and knock him out.

In June 1922 the noted English producer Charles B. Cochran was faced with an unexpected emergency. His latest revue, *Mayfair to Montmartre*, was forced to close because leading lady Alice Delysia had taken ill, and he was desperately in need of another attraction to keep his theatre open. A friend had just seen Clark and McCullough in New York in a burlesque show called *Chuckles of 1922*, and, at his urging and sight unseen, Cochran imported the entire revue to London. Thus it came to pass that a burlesque show considered below the artistic standards of the Broadway legitimate theatre was being sponsored without alteration by London's most distinguished impresario. But *Chuckles*—or more precisely Bobby Clark—scored such a hit that the show could have remained far longer than the two months it played the West End. The reason for the limited run: Irving Berlin, having seen *Chuckles* in New York, had already signed Clark and McCullough to appear on Broadway in his *Music Box Revue* in October.

The Berlin show, the second of four *Music Box Revues* presented by Sam H. Harris at the Music Box Theatre, marked Clark and McCullough's emergence as a major comedy team. Sharing the bill with such attractions as long-legged comedienne Charlotte Greenwood, stalwart tenor John Steel, and the harmonizing McCarthy Sisters, the burlesque duo easily fitted into the more decorative trappings of one of the most tasteful, imaginative, and artistic revues ever presented on Broadway. As Heywood Broun commented in the *World*. "The new surroundings, the silks and satins and the grandeur of it all abash Mr. Clark not the least. He finds it just as easy to knock them out of $11 seats* as in the old dollar days."

Clark and McCullough's best-received sketch, "The Lady in Red," had first been performed the previous year in the Lambs Club annual *Lambs Gambol*. A takeoff on spy-chase melodramas, the humor was derived by showing the production as it might be performed by a third-rate stock company in Winniepasooga, Wisconsin. Not only are the actors shaky in their lines, but the company's newly hired property man turns out to be totally inept. The lady of the title appears dressed in green. A calendar clearly shows the month to be January, even though the characters in the story comment on the oppressive heat. The grandfather clock strikes six when the hour is supposed to be midnight. Exit doors fail to open. This kind of manic humor was made to order for Clark appearing as Mohamed Mahoney, an arch conspirator, with McCullough as his henchman.

Two years later, the team was again signed for a *Music Box Revue*, this time appearing with Fanny Brice, soprano Grace Moore, and tenor Oscar Shaw. For this show the boys recycled two of their vaudeville and burlesque sketches—the prizefighter bit and a revised version of the lion-tamer skit with the animal now changed to a bear. For the first time Clark acted in two sketches without McCullough. In "The King's Gal" he played lascivious Lous XV to Miss Brice's seductive Mme. Pompadour, and in "Adam and Eve" he played lascivious Adam to Miss Brice's seductive Eve.

In 1923 producer Philip Goodman took a mumbling, bulbous-nosed former juggler and revue comic named W. C. Fields and elevated him to stardom in the musical comedy *Poppy*. The show ran almost a year. Looking around for other likely comic-star material, Goodman chose Clark and McCullough to appear in *The Ramblers*, their first book musical. The loose-jointed vehicle was written by Bert Kalmar and Harry Ruby, who had already provided Clark with two sketches in the 1924 *Music Box Revue*.

In the musical, Clark and McCullough played a spiritualist medium and his assistant. Stumbling on a movie company filming in Tia Juana, they help rescue the heroine after she has been kidnapped, a situation that involves them in mistaken identity, wild chases, and general nonstop activity. Brooks Atkinson, always one of Clark's most devoted admirers, welcomed the team in the *New York Times* with these words: "Here they are again, up to new as well as old antics, dressed elaborately in loose-fitting cloaks and variegated hats, passing gags back and forth at high pitch and winking confidently at the audience. Well swollen with bravado, Mr. Clark throws back his shoulders in his baggy topcoat, grins broadly, puffs his cigar rapidly, spits it out, snaps it

*Opening-night orchestra seats were double the normal ticket price of $5.50.

back in, uses both ends interchangeably, and swings his stick recklessly. Perfect as Mr. Clark may be as a low comedian, Mr. McCullough, in the fur coat and straw hat, matches him as a feeder, with a nervously grave sort of merriment."

The show's standout comic scene occurred toward the end of the evening. Clark and McCullough, the former decked out in a coonskin hat with a blanket rolled over his shoulder and the latter in a badly matched hunting outfit, take leave of the movie actors, determined to repair to a peaceful backwoods locale despite protestations that they remain. Suddenly, the emotion of the occasion becomes too much to contain in spoken dialogue, and the boys, almost to their own surprise, burst into the florid sentiments of Tosti's "Farewell." Booming "Goodbye, forever!" again and again, they get carried away by their operatic pretensions. Gesticulating broadly, they march up and down the stage shaking hands with their friends, then disappear into the wings only to reappear from the opposite side of the stage to thunder again their heartfelt goodbyes. Finally, still singing, they jump down from the stage to pump the hands and pat the backs of the more accessible members of the audience. As Walter Winchell described the scene in the *Graphic*, "How the crowd yelled, howled, screamed, rocked, and doubled up! It was the most demonstrative theatre attendance that has collected in one sitting in a long spell."

Though *The Ramblers* established the partners' theatrical legitimacy, the boys did not immediately return to the boards after the show's eight-month run. Instead, challenged by the new medium of sound film and dazzled by Hollywood offers, they spent 1928 and 1929 making a series of fourteen film shorts for Fox. But Clark and McCullough found little satisfaction in their West Coast experience and were happy to return to New York for what was to be the most notable musical production they ever acted in as a team.

In 1927 George and Ira Gershwin had joined librettist George S. Kaufman to collaborate on a stinging anti-war musical satire called *Strike Up the Band.* Coming only eight years after the end of World War I, its tale of the United States becoming involved in a war with Switzerland over the tariff on cheese—and ending with preparations for a war with the Soviet Union—found audiences so unreceptive that the show folded in Philadelphia before reaching New York. Undaunted, producer Edgar Selwyn prevailed upon playwright Morrie Ryskind to alter Kaufman's libretto to make it more commercially palatable. The disputed product was changed from cheese to chocolate, the battle scenes were put within the framework of a dream, eight songs were added for eight that were dropped, and the entire cast was replaced. Now co-starring in *Strike Up the Band* when it opened on Broadway early in 1930 were Clark and McCullough with Bobby in the enlarged role of Colonel Holmes (vaguely suggesting President Wilson's adviser, Colonel House) and Paul in the newly created part of Holmes's assistant.

With its lighter, more conventional tone, the show garnered generally favorable reviews and went on to a run of almost 200 performances. The casting of two broad comics in the starring roles not only gave the musical some needed swift-paced clowning but to a large extent established the kind of individual Clark would play in most of his appearances in book musicals—a scampering, beaming, mischievous, concupiscent zany, totally unabashed by being placed in a position of authority and respect. The character may be something of a scalawag, but at all times he is someone who never takes himself or his predicament seriously. He may be out of place and out of his depth, but he is always comfortable in his surroundings. No one can make him feel embarrassed because he is simply too busy enjoying his own wild and wicked behavior.

Of the eight new songs in *Strike Up the Band,* two of them—"Mademoiselle in New Rochelle" and "If I Become the President"—were created specially for Bobby Clark. The first was a lively if unmemorable number for him and McCullough to perform while cavorting with Swiss maidens, which Clark interrupted with his by now familiar disclaimer of vocal skill, "We don't sing good but we sing loud." The second song, a cleverly rhymed showstopper, was both a marriage proposal to wealthy dowager Blanche Ring and a forecast of their life together as the nation's First Couple. During the Boston tryout, Clark wrote four pages of additional lyrics that he wanted to sing as encores. Though appalled at the comic's audacity—as well as the quality of his contributions—lyricist Ira Gershwin was persuaded to go along with some of his additions. According to the perplexed Gershwin, a Bobby Clark interpolation such as "And just to show we're home-like/ We'll bathe in the Potomac," was greeted with as many laughs as any of the more skillfully crafted lines he had written.

The aim of *Strike Up the Band* may have been Gilbert and Sullivan satire, but the show did not lack for its share of traditional musical-comedy romance and dancing girls. Or traditional Clark and McCullough. Here, however, their familiar entrance had a few new tricks. As Brooks Atkinson described it, "Clark nimbly fingers his cigar, does a rapid manual of arms with his stick, blows opaque puffs of cigar smoke inside his coat and recoils with horror when stray wisps appear below. When the cigar drops on the stage, like alert vagrants and true, both he and McCullough pounce for it."

The season following *Strike Up the Band* Clark and McCullough returned in *Here Goes the Bride,* a concoction by cartoonist Peter Arno that lasted a week. Clark played a valet and McCullough the valet's valet, both in attendance on two couples who go to Reno to straighten out a lovers' quadrangle. By this time Atkinson had become so totally devoted to Bobby Clark that even when his material was threadbare he could write, "Now that the Marx boys have risen to the lacquered splendors of the screen, Clark and McCullough are the logical First Actors of the stage. Mr. Clark is a humorist as well as a clown. After anxiously applying a stethoscope to the paunch of the fat man in the show, he gives a really learned diagnosis: 'He has mice.' To those who understand the subtleties of art, Bobby Clark is a hero."

In 1932, Clark and McCullough were co-starred with the celebrated Beatrice Lillie in *Walk a Little Faster,* a revue that attempted to follow in the sophisticated paths of such entertainments as *Three's a Crowd, The Band Wagon,* and *Flying Colors.* But despite the talents involved, the show (whose title suggests that originally it may have had some kinship with *Alice in Wonderland*) turned out to be a rather humorless affair. Still, the Clark and McCullough entrance bit did have a new twist. "Remember, Paul," Clark advised his partner, "This is a classy new show. We can't pull any of that old stuff." And with that he once more spat out his cigar and rapped McCullough on the knuckles when he tried to pick it up.

S. J. Perelman contributed two sketches, neither memorable. In "Scamp of the Campus," set in a girls' college in 1906, Miss Lillie played temptress Penelope Goldfarb with Bobby as Sport Cardini, her Latin lover. In "Moscow Merry-Go-Round," inspired by the Russian purge trials, Bobby assumed the unlikely guise of Dictator Stalin as he takes direct action to cut down the size of the Politburo. He also played Sourdough, a miner, to Miss Lillie's Frisco Fanny in a sketch set in a Yukon saloon, and he mimicked dancer Clifton Webb's suavely romantic movements in another routine with Miss Lillie.* *Walk a Little Faster* kept strolling for about three months.

Thumbs Up!, the second Clark and McCullough revue in a row, came to New York at the end of 1934. Lavishly produced by Eddie Dowling, who was also in the show, and directed by veteran John Murray Anderson, the cast of sixty also featured madcap comic Ray Dooley (Mrs. Dowling), vaudevillian Rose King, the harmonizing Pickens Sisters, impersonators Eddie Garr and Sheila Barrett, and dancers Hal LeRoy, Paul Draper, and Jack Cole.† Though Clark and McCullough were well received, it was Miss Dooley who won the bulk of critical and popular acclaim, especially for two uproarious sketches. As a member of a troupe of Arab acrobats she was ruthlessly tossed about before turning

*Bobby was the second comedian that season to do a takeoff of Webb, who was then appearing in *Flying Colors.* An even more devastating impression was Bert Lahr's in *George White's Music Hall Revue.*

†Following the opening, *Variety*'s critic, Abel Green, observed, "That $4.40 top is an item which may prove a hurdle. A $3.30 scale, considering the St. James Theatre's generous capacity, may be a future compromise for the better."

into a jellied mass of fear as she shakily clambered to the top of a human pyramid. Later, in a *Merry Widow* spoof, she danced dizzily with her lover, Bobby Clark, then hiked up her skirt and scurried over the bowed backs of her ornately caparisoned guardsmen. Clark's funniest sketch was "Aired in Court," in which he played the microphone-hogging bean-shooting judge at a murder trial that was being broadcast. "Give Mr. Clark a gavel and the ribs are his to crack as he pleases," wrote John Anderson in the *Journal*. "With his painted-on glasses and that superior glitter in his eye he administers the microphone and lets justice fall where it may, which is usually in a heap."

For the first time on Broadway, Clark and McCullough did away with their trademark opening bit, the struggle over Bobby's dropped cigar butt. *Thumbs Up!*, however, presaged an even far more significant development in their relationship. Even though Clark and McCullough were featured as a team, the latter appeared in only two of the eight scenes in which Clark took part. Sadly, he was hardly missed.

Clark and McCullough made twenty-one film shorts for RKO between 1931 and 1935, and toured in the *Earl Carroll Vanities* early in 1936. Still, there was something clearly amiss about the partnership. McCullough, who was becoming increasingly depressed about devoting his career to being another man's stooge, now had to face the humiliating fact that even his secondary position in the act was no longer essential to its success. After the team's stint in the touring *Vanities*, McCullough entered a Massachusetts sanitarium suffering from nervous exhaustion. On his way back to New York after his release, he stopped off at a barbershop for a shave; as he was leaving he picked up the barber's razor and slit his throat and wrists.

Following his partner's horrifying death, the distraught Clark cancelled all bookings and went into seclusion for several months. He came back in the fall of 1936 to appear with Fanny Brice in the second edition of that year's Shubert-produced *Ziegfeld Follies*. Except for cast changes, it was a near replica of the first edition, which had been terminated because of Miss Brice's illness. Now with Bobby taking over roles that had previously been played by Bob Hope and Hugh O'Connell, the revue was greeted by the press as a funnier, tighter, and stronger production thanks chiefly to the combination of Brice and Clark. As John Mason Brown put it in the *Post*, "When seen singly, Miss Brice and Mr. Clark are enough to satisfy the greediest of playgoers. But when seen together on one bill, and when both of them are in topnotch form, the causes for gratitude increase by arithmetical progression, especially when surrounded by such a workmanlike, well-danced, vastly improved and only occasionally nodding revue."

In this edition, Bobby Clark sang "I Can't Get Started with You" to Gypsy Rose Lee, played the hero in the Hollywood satire, "The Gazooka," was Fanny

Brice's hysterical husband in "The Sweepstakes Ticket," was seen as the exasperated Hollywood director in the sketch about Baby Snooks, appeared as Roosevelt brain-truster Rexford Tugwell (here called Rexford Givewell) in a scene that poked fun at the New Deal's excessive generosity, and, in "Fancy Fancy," acted the role of Miss Brice's titled English lover. And to prove that all was still basically right in the Bobby Clark world, the comedian made his entrance grappling with a chorine over his dropped cigar. The second *Follies* of 1936—with the top price reduced from $5.50 to $3.85—ran as long as the first and then toured.

For such an individualist as Bobby Clark to be able to portray characters in sketches that had been created for other actors proved the breadth of his remarkable talent. A test of a different kind came along in 1939. The previous year, the raucous, near-sadistic vaudeville team of Olsen and Johnson had scored an unexpected smash with their knockabout, freewheeling show, *Hellzapoppin.* Now in a similarly unbridled spirit they presented—but did not appear in—*The Streets of Paris.* To make sure that the revue's comedy would be not only fast and low but also plentiful, they signed both Bobby Clark and the team of Abbott and Costello, burlesque graduates who had recently won fame on the Kate Smith radio program.

Despite competition from two newcomers (as well as the dynamic Carmen Miranda, here making an auspicious North American debut), Bobby's contributions were by no means overshadowed. In a scene set in a Montmartre dive, he and Della Lind sang the love song "Is It Possible?" with such single-minded concentration that they were totally oblivious of the mayhem around them caused by murderous Apaches throwing knives, chairs, and each other into the air. Even a stageful of corpses and a dagger in the seat of his pants could not distract Bobby from his ardent warbling. In another sketch, "The Convict's Return" (by Frank Eyton), Bobby assumed the protean task of playing four different characters who, since they could not appear on stage at the same time, made it necessary for a number of speedy backstage costume and makeup changes. First Bobby shows up as the senile father and then as the faithful butler of unhappy Marie (played by Luella Gear), who is pining away for her falsely imprisoned fiancé, Armand. Suddenly, the escaped prisoner (Bobby Clark) bursts into the room, followed by the hotly pursuing warden (Bobby Clark). The humor develops from Bobby's increasing difficulty in making his changes quickly enough (conveyed to the audience by the amplified thump-thump-thump of his frantic dashing from side to side behind the set), which is aggravated by his frustration with doors that keep jamming, windows that keep slamming on his head, and drapes that keep blocking his movement. By the time the sketch is over, with Armand in Marie's arms after a pardon is received from the governor, Bobby is so exhausted that he collapses on the floor.

The Streets of Paris also gave Bobby his most memorable song, "Robert the Roué from Reading PA" (by Jimmy McHugh and Al Dubin). With eyes flashing with roguish delight, Bobby sang it while strutting up and down the stage, twirling his cane and puffing his cigar, as he confided to the audience the amorous conquests of a tourist in Paris ("And I usually play in the hay—Hey!").

Nothing Bobby Clark had ever done before quite prepared audiences for his first appearance in a non-musical production. Ever since 1922, The Players, New York's venerable theatrical club, had maintained a tradition of reviving classics with all-star casts and presenting them on Broadway for limited one-week runs. Though the series had ceased in 1936, four years later the club staged a special revival of *Love for Love,* William Congreve's 245-year-old comedy of manners and morals. In the cast were to be found such distinguished actors as Cornelia Otis Skinner, Dudley Digges, Dorothy Gish, Peggy Wood, Romney Brent, and the former star of the Columbia Burlesque Wheel—Bobby Clark. Bobby played the role of Ben Legend, the lusty, rough-talking seafaring son of Sir Sampson Legend, who has returned to London to collect an inheritance, win a bride, and, in general, disrupt the artificial world of mincing dandies and their elegant ladies. In short, except for appearing without his "glasses" for the first time, Bobby had a part that let him behave much as he usually did in musical comedy. It even gave him a chance to be seen in a skirted sailor uniform and to sing—and do a hornpipe to—a bawdy ballad about the indiscriminate amours of "Buxom Joan of Deptford."

Bobby Clark's dominance of the production sparked interest in casting him in other classical roles. Just a few months later he signed with the Theatre Guild to appear as Feste in *Twelfth Night,* co-starring Helen Hayes and Maurice Evans. During rehearsals, however, he left the cast because of disagreements with the director over his interpretation.

Easily the most forgotten Bobby Clark appearance on Broadway was in *All Men Are Alike* in 1941, his only non-musical with a modern setting. (He was to have been in William Saroyan's *The Cave Dwellers* in 1957 but withdrew because of ill health.) *All Men Are Alike* was a featherbrained British farce that awkwardly combined German spies, philandering husbands, irate wives, and Colonel Blimpish army officers. Again minus his "glasses," Bobby resorted to all kinds of tricks to pump humor into the offering. He dashed through banging doors, swung out of a window on a hanging vine, paraded around in his wife's wartime uniform, hopped up and down a flight of stairs wrapped in a thick rug, and—in his most heavily applauded bit—kicked a pair of trousers into the air so that it would land with the suspenders looped over the staircase newel post. "He is more than a clown," asserted John Mason Brown, "he is a circus in himself." Brooks Atkinson, however, was not one for such restraint. "He is funnier than a circus," he wrote. "If all the clowns were put together, Bobby could outmatch them laugh for laugh and toss in a couple of side shows for good measure."

Having met and mastered William Congreve, Bobby next set his sights on Richard Brinsley Sheridan. His return to classical comedy was in the role of Bob Acres in *The Rivals*, which opened in January 1942. Co-starred were Mary Boland as Mrs. Malaprop and Walter Hampden as Sir Anthony Absolute. Eva Le Gallienne, who directed the production for the Theatre Guild, admitted that it didn't take her long to realize that the only way to direct Bobby Clark was to let him direct himself. In fact, when he showed up for the first rehearsal, the comedian had already covered his script with notations indicating where he planned to indulge in such specialties as handsprings, headstands, whip cracks, and costume changes. And he never stopped improvising until opening night.

In the most comical scene in the play, Bobby is egged on by Sir Lucius O'Trigger to write an indignant letter challenging a rival to a duel. Trying to muster courage to pen so foolhardy a message, Bobby seats himself at an imposing desk and carefully chooses a quill. When the quill somehow doesn't respond to his scratching, he angrily sends it flying across the stage. Other quills are tested and are dispatched in the same fashion. Then Bobby turns on Sir Lucius, expressing in growls his irritation for providing such defective writing equipment. Back to the letter, Bobby gets himself into a properly bellicose mood through a variety of bravado looks and poses until, having finally found a quill that works, he manages with deliberate effort to write one word. After viewing it with pride, he briskly sprinkles it with an excess of sand, then blows the sand away in a cloud of dust, bangs several desk drawers, and scampers up to the top of a huge wing chair. There he finds enough creative inspiration to scratch another word, followed by more sand sprinkling, more blowing, and more deliberation. Eventually, he finishes the letter with his feet hooked over the back of the chair, his body suspended in the air, and his chin resting upon the desk.

During the performance Bobby also found a spot in which to sing again about "Buxom Joan of Deptford," thus linking the role of Bob Acres with that of Ben Legend in *Love for Love*. They both, of course, possessed aspects of the familiar Clark personality, especially in their totally unselfconscious zaniness, but it was the role in the Sheridan play that Bobby claimed as his all-time favorite. "Poor little country bumpkin," he once remarked to an interviewer. "I felt exactly as if Sheridan had me in mind when he wrote the part."

From *The Rivals* to *As the Girls Go* almost seven years later, Bobby Clark ruled supreme as the Clown King of Broadway both in classical comedies and in musical comedies, adjusting his characterizations without readjusting his style or his technique. In 1942, in fact, he could be seen in the Theatre Guild's *The Rivals* (in which he also toured) as well as in Michael Todd's glorified burlesque show, *Star and Garter* (which opened in June and kept running until December 1943). Hassard Short was the director.

Bobby's first scene in the revue was in a sketch he wrote himself called "The Merry Wife of Windsor." In it he is found enjoying the favors of the merry wife, played by Gypsy Rose Lee, when her husband returns home unexpectedly. Forced to hide, Bobby is about to climb into a trunk when he stops in his tracks, turns to the audience, and moans, "Why did I ever leave the Theatre Guild?" The show's success turned out to be reason enough.

Unfortunately, much of the material was warmed over, including Bobby's "Robert the Roué" number from *The Streets of Paris,* the skit about the microphone-hogging judge from *Thumbs Up!,* and a revised version of Abbott and Costello's "Rest Cure" sketch in *The Streets of Paris,* here called "Crazy House." One new scene, "In the Malamute Saloon," offered a sampling of the way burlesque humor was being treated on Broadway. Bobby, as a grizzled miner, staggers into the saloon and asks the bartender the way to the powder room. When the bartender points outside, Bobby opens the door and is greeted by a faceful of snow. Closing the door, he announces matter-of-factly, "It's too cold outside. I'll wait until spring."

Within weeks after closing in *Star and Garter,* Bobby Clark was again starring on Broadway in a Mike Todd production directed by Hassard Short. *Mexican Hayride* had originally been conceived with William Gaxton and Victor Moore in mind, but their financial demands were too high. The script was then rewritten to fit the role intended for Moore to the specifications of Bobby Clark. An even more resplendent production than *Star and Garter* and one of the most elaborate of that wartime period, *Mexican Hayride* was Clark's first book musical since *Here Goes the Bride.* It was also his only association with composer-lyricist Cole Porter. Porter wrote a number of songs for Bobby, who ended up doing only two. "Girls," performed with a bevy of admiring lovelies, was a spirited piece which, like "Robert the Roué," gave him the chance to boast of his amatory appeal, and "Count Your Blessings" was a showstopper he bellowed with June Havoc and George Givot in which the miseries of life were deemed to be at least preferable to death.

Clark's comedy, as always, was not dependent upon songs or script (the work of Dorothy and Herbert Fields); in fact, well into the show's run Bobby was still introducing new jokes and business. As before he wore his painted-on glasses, gargled his "rrrr's," and skittered madly about, but some of the new routines were little short of inspirational. As a numbers racketeer on the lam in Mexico who is accidentally acclaimed the "Amigo Americano" as a goodwill gesture, Bobby played an out-and-out swindler for the only time in his career. Alternately hailed by the populace and trailed by the police, he disguised himself as a Mariachi flute player, and later masqueraded as a cigar-chomping squaw vending tortillas, enchiladas, and tamales. For some reason, he decided that what he needed most was a mouthful of protruding teeth and, ever the perfectionist, he had a dentist make them to his specifications. To round off his outlandish appearance, he added one master touch: a doll papoose made as a miniature version of himself—complete with "glasses," buck teeth, and tiny cigar—which he had strapped to his back. One of the cherished sights in this getup came when Clark inhaled the cigar and the papoose (by means of a concealed atomizer) then seemed to be exhaling the smoke.

In 1946 Bobby Clark returned to the classics. With the aid of William Roos, an uncredited but experienced playwright, he fashioned his own adaptation of Molière's *Le Bourgeois Gentilhomme,* a satire on the French social scene of 1670. In this somewhat free version—titled *The Would-Be Gentleman*—he was less an actor playing a role than a star performer doing a series of farcical routines. The play centers upon the nouveau-riche Monsieur Jourdain who, anxious to become accepted by Paris society in order to win a titled mistress, has a series of tutors instruct him in dancing, music, fencing, and philosophy. They, of course, play upon his gullibility by teaching him only the most rudimentary of skills and the most obvious of information (the philosophy instructor wins Jourdain's admiration by solemnly advising him that he has been speaking prose all his life without knowing it).

It is easy to understand why this character would appeal to Bobby Clark. Again he could be the unschooled but uninhibited parvenu, with even greater latitude than before to adjust the character and situations to his own comic style, taking full advantage of props such as foppish wigs, snuff (which he called "snoof"), brocade coats, and dress swords. (When he first brought the script to producer Michael Todd, he put a notation in the margin: "At this point, I throw a snuffbox across the stage and it falls into the valet's pocket. This can be done. I have been practicing it.")

Combining the Molière rapier with the Clark broadsword left no doubt that satire would be subverted to slapstick. Everything was broadened and coarsened, yet the star's energy and inventiveness proved indispensable in infusing new life into the tale. Bobby's brand of comic resuscitation was also much needed when, in 1947, he starred in a revival of the thirty-four-year-old Victor Herbert operetta *Sweethearts.* For the first time he was in the middle of mythical-kingdom romance, playing the role of Mikel Mikeloviz, a political operator involved in all sorts of complicated intrigue concerning a royal foundling who has been adopted by a laundress and ends up as a queen. Or as Bobby put it in an ad-lib aside to the audience, "I'll convince Dame Lucy that I'm her long-lost husband, find out which girl is the adopted daughter, and become the big shot of Zilania—and that's all of the plot you'll get out of *me!*"

Sweethearts even found Bobby looking as he did in the days of Clark and McCullough. Not only were his painted-on eyeglasses, cigar, and cane restored, he even made his whirlwind entrance wearing his once familiar porkpie hat and short topcoat. As might be expected, the musical was little more than an excuse for a Bobby Clark romp in which, no matter the demands of the plot, he happily inhabited a world of his own. After making a comment to himself while alone on the stage, he apologized to the audience, "Pardon me for talking to myself. It's an old operetta custom." He disguised himself as a washerwoman and deftly heaved a union suit into the air and made it land perfectly on a rooftop clothesline twenty feet away. He scrubbed his teeth with cigar ashes. He interrupted a romantic duet to appear with a two-wheel cart and inquire across the footlights, "Has anyone seen a horse run down the aisle?" He performed a headstand on a sofa. He change for a no apparent reason into a dozen outlandish costumes, including those of a French Foreign Legionnaire and a Dutchman with wooden shoes. During a barely intelligible story development, he confided to the audience, "Never before has a thin plot been so complicated." Leading a sextet of monks, his head covered with a ratty wig, he caterwauled "Pilgrim of Love" with such unabashed bravura that he turned it into the musical highlight of the evening. As Richard Watts, Jr., observed in the *Post*, "He superbly demonstrates his rare talent for being at the same time leering and innocent, ferocious and sweet, energetic and effortless, while he goes about the tremendous business of making a great deal out of nothing at all with frantic nonchalance."

There seems little doubt that without this irreverent clown, the musty, fustian *Sweethearts* never would have had a chance on Broadway in 1947. Bobby, however, kept the customers howling for almost 300 performances before he took the show on the road.

As the Girls Go, the following year, was Bobby's fourth association with Mike Todd. It also marked his final appearance on Broadway. Though the musical's 420-performance run put it right up there with such previous Clark–Todd hits as *Star and Garter* and *Mexican Hayride,* it succeeded only after having been beaten and pummelled into shape during its Boston tryout.

The show was based on a premise far less outlandish today than it was in 1948: the election of the first woman President of the United States. Irene Rich, primarily a film and radio actress, played the Chief Executive and Bobby was her husband, the First Gentleman of the Land. Initially, the concept was to have been a satire in the *Of Thee I Sing* tradition, with the humor far more cerebral than in any musical Clark had ever done before. No broad comedy, no jokes that weren't part of the theme, and no songs and dances that did not move the story forward or convey the proper satirical viewpoint. Bobby wasn't even allowed to paint on his glasses.

The result was that the Boston critics pronounced the show a disaster, with the *Variety* representative leading off his review with a note to the producer: "Dear Mike, Close it and forget it." But Todd had too much at stake to heed the advice and, with only one week to go before the New York opening, he scrapped most of the book and put the emphasis on the kind of rowdy, high-spirited entertainment that had proved so successful in the past for both Clark and himself. Back went Bobby's "glasses" and with them a free hand to devise whatever comic business the comedian could dream up. Anything that got a laugh stayed in; anything that held up the action went out. Bobby chased Amazonian show girls, blew soap bubbles out of a bugle as he led a bugle-tooting band of youngsters, impersonated a manicurist and a lady barber (with crumpled towels for a bosom and a horse's tail for a hairdress), wore a huge fur coat and carried a matching fur cane, tossed sugar lumps unerringly into a distant teapot, and, without looking, hurled his hat across the stage and made it land on a hatrack. In one scene, Bobby, obviously on his way to take a trip, dashed on stage with two porters carrying snow shoes, a diver's helmet, and a golf bag containing a pair of skis. When reporters asked him where he was heading, Bobby flicked his cigar and brusquely replied, "I don't know—as you can plainly see."

To everyone's amazement, somehow it worked. *As the Girls Go* was welcomed enthusiastically by almost all the Broadway reviewers, and even at the unprecedented top ticket price of $7.20, Bobby kept the customers laughing for thirteen months. His final performance on Broadway was on January 14, 1950.

This was not, however, Bobby's farewell to the theatre. Just five months later he was again associated with a Michael Todd burlesque-type revue, *Michael Todd's Peep Show,* but only as the director of the sketches (which found him, for reasons none too clear, billed as "Robert Edwin Clark, Esq."). In 1952, Bobby was to have returned to Broadway in a revised version of the Sammy Fain–E.Y. Harburg musical, *Flahooley,* which had had a brief run in the spring of 1951. Retitled *Jollyanna,* this social satire about the invention of a laughing doll was tried out in San Francisco and Los Angeles under the auspices of Edwin Lester's Civic Light Opera Association. Bobby, who appeared in a greatly altered variation of the toy manufacturer role that Ernest Truex had played on Broadway, was co-starred with Mitzi Gaynor. But despite changes, the show was still weighted down by a socially conscious book, and it never made it to New York.

Bobby Clark's last stage appearance was in the touring company of *Damn Yankees*, which kept him on the road between January 1956 and May 1957. His part was that of the devilish Mr. Applegate, a role originated by Ray Walston, who transforms the middle-aged baseball-loving hero into a youthful homerun king. But conveying evil—no matter how comically conceived—was inimical to Bobby's basically naïve personality. To give him the properly Satanic look, he penciled in two arched eyebrows and even, for a time, tried doing without the greasepaint glasses. His appearance in the musical was an event of such importance to Brooks Atkinson that the *Times* critic filed a special report from Washington, D.C., soon after the tour began. In it he wrote, "Bobby has scaled himself down to the dimensions of the book without losing the impudence and gusto. Teetering perilously between the sublime and the ridiculous, he is entertaining the public, as he always has, with the skills of a great mountebank who has never lost the innocence of his approach to comedy situations."

Bobby Clark died in New York on February 12, 1960, at the age of 71. Though Broadway had not seen him in ten years, he was the last of that special breed of comedians whose fame was achieved exclusively in the legitimate theatre. What is remarkable is that his appeal was never stronger than it was during a period that saw some of the most influential changes within the style and substance of the musical stage itself, changes that seemed to doom the very theatre in which Bobby Clark excelled. No greater tribute could be paid this durable clown than to note that his last three Broadway appearances, when he was at his peak of popularity, took place during the same period as *Oklahoma!*, *Bloomer Girl*, *Carousel*, *Finian's Rainbow*, *Brigadoon*, *Kiss Me, Kate*, and *South Pacific*.

Joe Cook

"I WILL GIVE an imitation of three Hawaiians. This is one (whistles). This is another (tinkles mandolin). This is the third (marks time with foot). I could imitate four Hawaiians just as easily but I will tell you the reason why I don't do it. You see, I bought a horse for $50 and it turned out to be a running horse. I was offered $15,000 for him and I took it. I built a house with the $15,000, and when I was finished a neighbor offered me $100,000 for it. He said my house stood right where he wanted to dig a well. So I took the $100,000 to accommodate him. I invested the $100,000 in peanuts and that year there was a peanut famine so I sold the peanuts for $350,000. Now why should a man with $350,000 bother to imitate four Hawaiians?" (Exit.)

Joe Cook's imitation of four Hawaiians—or rather his reason for not imitating four Hawaiians—was one of the truly legendary routines of vaudeville and the musical stage. The above basic tale could be shortened or lengthened or altered or changed completely in favor of another story, depending upon Joe's mood or the audience's. The unalterable point was that no matter how outlandish, pointless or involved the reason might be, it always ended with Joe resolutely refusing to imitate more than three Hawaiians (which he always pronounced "Hy-wyans").

The monologue came into being during a vaudeville engagement in Akron, Ohio, in the mid-1910s, when Joe announced that he would imitate three Hawaiians at the same time. After whistling, playing his mandolin, and tapping his foot, he brashly told his audience, "I could imitate four Hawaiians, but I won't." Realizing that he would now have to give some kind of goofy reason for failing to go through with the imitation, he ad libbed, "Ladies and gentlemen, you have just seen me imitate three Hawaiians. Why should I do four Hawaiians and show up those who can only do three Hawaiians? Why should I be responsible for their losing their jobs?" And with that he walked off the stage.

Bland, cheerful, brash, ingratiating, glib, engaging, amiable, breezy—these were the adjectives usually applied to Joe Cook. His boyish round face was almost always wreathed in a wide, beaming, slightly smirky smile that could easily light up an entire stage. Through a clipped Midwestern twang he would spin rapid-fire rambling tales, mostly about his youth in Evansville, Indiana, or prattle on in double talk that could make the gullible listener believe anything he wanted him to. But no matter how hard he tried to play the conniving city slicker he was in reality an innocent, vulnerable rube with whom audiences could readily sympathize.

Among Cook's specialities were cumbersome, Rube Goldberg-type contraptions which he demonstrated with the look of genuine wonder and delight of a child showing off a sand castle. Though he was hardly a social satirist, basically these inventions were zany commentaries on an overly industrialized society in which machines of mind-boggling complexity are used to create products of absurd simplicity or uselessness. Additionally, Cook conceived of smaller labor and time-saving devices designed to enhance their owner's comfort which were similar to—if more functional than—the sight gags dreamed up by another inspired clown, Ed Wynn. But nothing either man ever did was to be taken seriously. Like Wynn, Joe Cook was supreme at purveying nonsense for the sheer unalloyed fun of it. He never burlesqued or ridiculed or spoofed. He simply reveled in pointless, complicated machinery the way he reveled in pointless, complicated monologues.

Since his more elaborate creations had to be constructed with the precision of a factory assembly line, Joe could not subject them to the kind of trial-and-error break-in of verbal routines. "If my sight gags are funny," he told an interviewer in 1940, "they are the result of the faith that moves mountains and actuates practical jokes. There isn't any justification or rationalizing or explaining them in advance, and I've never worked with a producer who had the slightest belief that any one of my scrammy brain children would ever get a laugh when actually produced. Most of the time I don't believe they'll work myself. I get the idea of an act, say, involving three dwarfs, a pile driver, a small scenic railway, and a gross of custard pies, and it seems funny as all getout when it first becomes my vision of the day. By the time I have secured the three dwarfs from a casting agency, and have commenced the structural work on the scenic railway, and contracted for the custard pies, the whole thing seems like a mirage. All of a sudden I feel it's bogus and won't work. Then on opening night I know it simply has to work or I'm all washed up with the cash-paying public, with the producer, and most of all with myself."

In addition to his smile, his manic inventions, and his non-sequitur stories, Joe was famed for his skill at a variety of circus specialties. W. C. Fields was as adept a juggler and Bobby Clark his equal at tumbling, but Joe Cook could also balance on a slack wire, ride a unicycle, lie on his back and twirl fellow performers in the air with his feet, walk a huge rubber ball up and down an incline, and perform a variety of other such tanbark feats. As Brooks Atkinson once wrote—with pardonable hyperbole—"Next to Leonardo da Vinci, Joe Cook is the most versatile man known to recorded times."

Leonardo's successor was born Joseph Lopez in 1890 in Evansville, Indiana (Joe was never sure of the exact date). His father, an artist of Spanish descent, drowned in a lake when Joe was three, and his mother died soon after from shock. Brought up by an Evansville farmer and his wife, whose name was Cook, Joe spent as many non-school hours as he could marveling at the feats of the acrobats and jugglers in the traveling tent shows. His ambition to emulate them was encouraged by his foster mother, who outfitted more than half of a large barn with all the necessary equipment. As Cook later recalled, "We had electric lights in the barn even before they were installed in the house."

Joe soon became so skillful that he gave performances in the barn (admission price five cents), and even before his teens he was touring the area in medicine shows. In 1906, at the age of fifteen, he was so convinced of his future as a performer that he left home to seek employment in New York. There Joe won his first engagement with the help of a doctored photograph showing him juggling seventeen balls in the air at the same time—which led to an early monologue about his inability to juggle seventeen balls in the air at the same time. After only three months in small-time variety theatres, he had made such strides that he was engaged to play the prestigious Hammerstein's Victoria, then New York's leading vaudeville house. Three years later, now a two-a-day headliner, he advertised his "One-Man Vaudeville Show" with the following message in *Variety:* "Master of All Trades. Introducing in a 15-minute act, juggling, unicycling, magic, hand balancing, ragtime piano and violin playing, dancing, globe rolling, wirewalking, talking, and cartooning. Something Original in Each Line—SOME ENTERTAINMENT!" In all, Joe toured the vaudeville circuits for fifteen years.

On October 6, 1919, billed for the only time as Joseph Cook, he made his Broadway debut at the Liberty Theatre in a revue called *Hitchy-Koo, 1919.* This was the third of four annual shows named for and starring a slack-jawed, sandy-haired comedian named Raymond Hitchcock. In addition to supporting Hitchcock in some of the comedy sketches, Joe offered a few of his vaudeville routines and introduced Cole Porter's song, "When I Had a Uniform On," a comic plaint of a demobilized soldier pining for the days when the girls found him irresistible in khaki. A few critics noticed Cook. According to the *Sun*'s reviewer, "One of the funniest parts is the scene in the steamship office in London. Joseph Cook as the clerk demonstrates his remarkable ability to in-

Joe Cook in his first starring hit, *Rain or Shine* (1928). *

A scene from Joe Cook's 1930 success, *Fine and Dandy*, with Dave Chasen giving his well-known hand wave.*

A handbill for *Fine and Dandy*.

dicate by nodding or shaking his head simultaneously 'yes' and 'no' at the same time." The show didn't last too long in New York, but it toured for a year.

In 1923 a third rival to the well-established series of annual *Ziegfeld Follies* and *George White's Scandals* was inaugurated by showman Earl Carroll. His offering, the *Earl Carroll Vanities*, which opened the Earl Carroll Theatre at 7th Avenue and 50th Street, was as elaborate as the *Follies* and as fast-moving as the *Scandals*, but Carroll went further than his rivals in exhibiting the undraped female form, and his sketches were far more censorable. To assure that his series began with appropriate publicity, he gave the leading female assignment to Margaret Upton Archer Hopkins Joyce, better known as Peggy Hopkins Joyce, whose romantic and marital involvements had made her front-page copy in the city's tabloids. Possibly to give the *Vanities* a contrasting air of innocence, Carrol made Joe Cook responsible for keeping the customers laughing.

Joe first appeared in the eleventh scene in the revue, listed on the program as "Joe Cook the Humorist, Presenting a Portion of his 'One-Man Vaudeville Show.' " In the scene he juggled balls and Indian clubs (he could keep six in the air at the same time), balanced on a Japanese pole, performed sleight-of-hand tricks, showed his skill as a marksman, plucked a banjo and blew a trumpet, walked a rubber globe up and down an incline, told a few rambling stories, and exited after giving his latest explanation of why he wouldn't imitate four Hawaiians.

The premiere *Vanities* was such a success for Earl Carroll and Joe Cook that the producer engaged the comedian for his next edition in 1924. This marked not only the first time that Joe was given feature billing but also the first time that he was joined by his faithful stooge Dave Chasen, who would appear in almost all of Joe's subsequent musicals. Chasen, who is best-remembered today as a Hollywood restaurateur, was then a stocky, curly-haired knockabout comic with an idiotic grin and one blackened tooth who reminded some people of Harpo Marx. His stiff palm-out hand wave—accompanied by popping eyes and that demented grin—became a much-copied gesture among theatregoers. The revue also revealed, in something called "The Electric Laboratory," the first of Joe's elaborate contraptions designed to inflict madcap punishment on Chasen's head. Later in the show Joe did what was expected: he didn't imitate four Hawaiians.

Joe Cook's initial Broadway book musical was supposed to be *How's the King?* in 1925, but it never got any closer than Philadelphia. Written by playwright Marc Connelly, with songs by Jay Gorney and Owen Murphy, it was about a multimillionaire playboy who buys a bankrupt Balkan kingdom just to become its ambassador to the United States and thus enjoy diplomatic immunity from the Prohibition laws. According to Connelly, Earl Carroll, the show's producer, tried cutting costs by hiring an inexpensive company to construct the scenery, with the result that the set for the first scene—the observation platform of Joe's private railroad train—collapsed opening night. No one was hurt, but nothing could save the show after that.

Since Cook was under contract to Carroll, the producer simply added him to the cast of the third edition of the *Earl Carroll Vanities*, then running on Broadway, and made enough changes in the songs, sketches, and cast to bill it thenceforth as the "Fourth Edition." Joe still found mileage in yet another variation on his Four Hawaiians routine, but he won his biggest laughs by reading and commenting on a story from *McGuffey's Second Reader* called "Who Stole Little Yellow Bird's Nest?" Here as always with a Joe Cook monologue, the humor was more in the telling than in the tale. In relating the account of the purloined nest, Joe not only became totally involved in the situation but also showed concern for the audience's reaction to the theft: departing from the text he tried to assuage worried listeners by observing that perhaps the nest wasn't really stolen but that Yellow Bird just had a hard time finding it among all those trees. As he went along, Joe enlivened the narrative by imitating animal sounds, such as Brown Cow's moo-moo and Black Sheep's baa-baa; then, looking up from the book, he commented that he sure hoped there weren't going to be any rhinoceroses in the tale. He also confessed that though he was burning with curiosity, he had had the will power to keep from sneaking a look at the final page to discover the thief's identity in order to share that discovery with his anxious listeners. With Cook heading the cast, this edition of the *Vanities* enjoyed a profitable cross-country tour during 1926.

The first Joe Cook book musical that did make it to Broadway was based on Joe's original concept about the travels and travails of a bankrupt circus troupe. *Rain or Shine*, which opened early in 1928, offered Joe in the well-tailored role of a circus manager, who gets his chance to perform a one-man show when the regular tanbark performers go on strike. To his feats of juggling, slack-wire balancing, rubber-ball walking, sharpshooting, and playing musical instruments, Joe now added knife-throwing, lion-taming, and banjo lessons (taught to a leopard).

Clad in a shako tilted to the side and an ankle-length coat with epaulets and a double row of brass buttons (but which was fastened by a zipper down the front), he also introduced his latest stage-wide mechanical invention, the Fuller Construction Company Recording Orchestra. This contraption was put into operation when Joe pulled a lever which started a whirling buzzsaw which goosed a man holding a soda-water siphon which squirted a man whose gyrations turned a Ferris Wheel whose five passengers took turns bopping Dave Chasen on the head with their violins. Each time he was hit, Chasen reacted with a look of childlike wonder as he tapped a triangle with a hammer, thus producing the teeniest tinkle to accompany Joe's trumpet playing of "Three O'Clock in the Morning."

With his first starring role on Broadway, Joe abandoned his Four Hawaiians routine in favor of an earnest, rapid-fire, totally pointless monologue about the joys of eating cornflakes with evaporated milk and the unhappy morning when he had to eat the cereal dry. What made audiences howl (while scratching their heads in bewilderment) was the brazen way Joe would stretch out his simple story and build it to a punchline—"It ain't good that way!"—that exploded with no punch at all.

In another scene, Joe, as "Smiley" Johnson, had this exchange with the saturnine, bespectacled Tom Howard, playing a gullible hayseed named Amos K. Shrewsberry:

AMOS: Ya know those carrot seeds ya sold me? Didn't grow.
SMILEY: Did ya plant 'em?
AMOS: Nope.
SMILEY: What did ya do with 'em?
AMOS: Ate 'em.
SMILEY: Out of curiosity?
AMOS: No, out of a plate.

The Cook conquest of Broadway was achieved without resistance. "Joe Cook looks like the average American fresh from a barber shop," wrote Alexander Woollcott in the *World.* "But behind that uneventful, almost ornery, façade there is a nimble, tirelessly inventive mind at work and just a flash of the most engaging lunacy, which breaks out from time to time in magnificent nonsense. Then he is an adroit comedian and one endowed with that hospitable quality which, at the moment when he steps cheerfully on stage, makes you suddenly glad you came to see the show." Tossing caution windward, Brooks Atkinson in the *Times* simply proclaimed Joe "the greatest man in the world," and went on to write, "Joe Cook is a comedian with the divine spark of madness . . . No doubt part of our enjoyment comes from the lack of ostentation in his nonsense. He seldom relies upon the bravura of carnival costuming. He does not smear his comedy with the greasepaint of a clown; usually, he spares the raucous slapstick of the low comedian. His style is glib, whether in patter or acrobatics. All the fortuitous transition of his monologues he manages with the swiftness and neatness of sleight of hand. For all you can tell, this god of fooling may be talking sense."*

*Joe Cook starred in a movie version of *Rain or Shine,* which Frank Capra directed for Columbia early in 1930. It was Joe's only feature-length film.

Rain or Shine's ten-month Broadway run made it Joe's biggest hit. It also established the basic pattern for his next two book musicals, *Fine and Dandy* and *Hold Your Horses.* Just as in *Rain or Shine* propelled Joe to the heights of circus clowndom, so *Fine and Dandy,* in 1930, traced his rise in business, and *Hold Your Horses,* in 1933, revealed how he became a winner in politics. All these shows, being tailored to his comic measurements, were sufficiently loose-fitting to provide plenty of room for wacky inventions and meandering monologues.

Fine and Dandy was the perfect Joe Cook carnival. In addition to the expected and essential ingredients it also offered a large and talented cast, colorful sets and costumes, and a superior score by Kay Swift and Paul James (a pseudonym for banker James Paul Warburg, Miss Swift's husband at the time). It even gave Joe the only song hit he ever introduced: the exuberant title number, which he performed with leading lady Alice Boulden (who would, a few years later, become Cook's second wife). The infectious piece ("Gee, it's all/ Fine and dandy/ Sugar candy/ When I've got you") won such sustained applause that the couple obligingly gave three encores with additional lyrics that found them assuming the roles of Amos 'n' Andy, Napoleon and Josephine, and heavyweight boxers Jack Sharkey and Max Schmeling. (In the last, Joe gave this advice to Alice on how to act the Schmeling part; "Clutch your vitals/ And claim six titles/ And take the boodle home with you.")

In the story, Joe played Joe Squibb, an easygoing worker at the Fordyce Drop Forge and Tool Factory. To impress his beloved that he does too have ambition, Joe—in a scene recalling Groucho Marx flirting with Margaret Dumont—is not above romancing the widowed Mrs. Fordyce, who owns the factory, in order to become general manager.

His first day as general manager finds Joe in his office with its huge window overlooking a very modernistic industrial plant. On one wall is a graph labeled "Production," and on another wall one labeled "Raw Materials." Since Joe finds these hard to understand, he busies himself by inspecting the contents of his filing cabinet. He opens one drawer and is faintly surprised to discover a large pair of men's shoes, which his embarrassed secretary (played by Eleanor Powell) claims to be hers. Another drawer produces a potted geranium. Still digging, Joe is a bit perplexed to find a barber pole. Then, with a slight shrug to indicate that nothing could possibly surprise him now, he pulls out a live skunk. Next a toothless old man walks in and, without explanation, munches a sandwich in a corner of the room. Seated behind an enormous desk, piled high with "In" and "Out" papers, Joe receives his first official visitor, Mr. Ellis, the former general manager who is also his girl friend's father. It also turns out that Mr. Ellis has embezzled company funds to cover his Wall Street losses, and he appeals to Joe to help him. "I am a big man," says Joe—and stands up on stilts.

At one of the many picnics that Joe gives to keep the workers happy, there is this typical Joe Cook exchange with a rich old lady known as Aunt Lucy, whom Ellis hopes will lend him money:

JOE: Mr. Ellis was telling me that you are interested in church work.

AUNT LUCY: That's not so.

JOE: Well, that's too bad. Possibly I got that confused with your interest in dumb animals.

AUNT LUCY: I detest dumb animals.

JOE: You're not interested in little kitties and doggies? You mean to say that you like to see little boys throw stones at birds? Do you think birds should throw little boys at stones?

AUNT LUCY: I don't like birds.

JOE: Neither do I. Shake on it. Didn't I read in the paper that you were donating a lot of money to a college?

AUNT LUCY: You did not.

And on and on . . .

Joe, in desperation, comes to realize that the only way to help Ellis is for them to rob a bank. Feigning casualness as he gives the bank the once-over, Joe is approached by a woman depositor with whom he carries on a conversation suggesting an early Woody Allen routine:

WOMAN: Aren't you Walter Simpson?

JOE: No. My name is Joe Squibb.

WOMAN: Well, you look exactly like Walter Simpson.

JOE: Oh, really?

WOMAN: I went to school with his sister. She's married now.

JOE: It's a small world, isn't it?

WOMAN: It certainly is. My, the good times we used to have together. Remember the night we all went down to the Mitchells?

JOE: Do I? Say, what a night that was.

WOMAN: I thought I'd die laughing. That was the night Fred Anderson fell off the porch swing. I'll certainly tell my husband I saw you.

JOE: Yes, be sure to say hello to him for me.

WOMAN: I certainly will. Are you around here much?

JOE: Yes, I'm in the book. Give me a ring.

WOMAN: I certainly will. Goodbye.

JOE: Goodbye. Stick 'em up, everybody!

Joe gets away with the attempted robbery because the bank president somehow has the notion that all he wants to do is to open a new account (as he advises Joe, "We give 2% on all balances over $1,000"), and they end up in a friendly discussion about the proper color for the checkbooks.

In addition to old tricks and such new ones as juggling cigarettes and balancing a tea set on his nose, Joe offered two more cumbersomely loony inventions. One machine was designed to crack walnuts, inflate paper bags with air so that they may be loudly deflated by hand, puncture toy balloons, and scratch one's back with a brush. It was also guaranteed to provide its owner with candy and flowers by having a mechanical fist knock out the would-be recipient and put him in the hospital. In the other, more elaborate creation, Joe again offered a performance by the Fuller Construction Company Recording Orchestra in which the living Rube Goldberg-type machine ended up producing the tinkle of a triangle. In this one, Joe blew a saxophone, which frightened a monkey in a tree into dropping a cocoanut on a jungle native who shot a missionary in the back with a bow and arrow which caused the engineer of a dredging machine to drop crockery and flour on the pate of Dave Chasen which somehow caused the single triangle tinkle.

And . . . oh, yes, the tale of the Four Hawaiians was restored.

Few performers and their vehicles have ever received such a warm welcome by the critics. Percy Hammond wrote in the *Herald Tribune:* "As the blue chip of American musical comedians, Joe Cook is unexelled in the variety and skill of his accomplishments. He turns handsprings, juggles flambeaus and plays the saxophone with the same quiet ease in which he cracks his jokes . . . *Fine and Dandy* is a laughing, handsome and respectable entertainment." In the *Times,* Brooks Atkinson observed, "Joe's engaging smile keeps everything delightful. To say that he would be a small boy's hero is merely to acknowledge the degree to which he delights grown men."

According to John Mason Brown in the *Post, Fine and Dandy* was "Not just a good show, but a grand and glorious one . . . one of the best musical comedies New York has seen in many a blue moon," and its star "fluent and inexhaustible as ever, and as gloriously nonsensical as only he can be." Topping the Brown appraisal, Robert Littell, in the *World,* called the musical "one of the best shows I have ever seen," and welcomed Joe as a riot-creating comet. "I do not, really, see what more any one can want of a musical comedy," opined the *Sun*'s Richard Lockridge, who termed it "an evening of the insane, nightmarish, extraordinarily hilarious irrelevencies which make this priceless comedian what he is." Robert Benchley in the *New Yorker* found the show "the funniest musical play in town." Gilbert Gabriel in the *American* hailed it as "the season's finest and dandiest fun . . . pretty nearly everything you've yearned for in the way of 1930 entertainment." And from Walter Winchell in the *Graphic:* "Joe Cook is certainly one of the musical theatre's three geniuses. I can't at the moment think of the other two."

When *Fine and Dandy* closed at Erlanger's Theatre on May 2, 1931, it had achieved a run of 255 performances, almost as long as such other hits of the same Depression season as *Girl Crazy, Three's a Crowd,* and *The Band Wagon.* It was also, sadly, Joe Cook's last great success. In 1932, director-producer John Murray Anderson had the notion that Joe's brand of comedy would go over well in London, and he signed the comedian to appear in a West End revue called *Fanfare.* According to Anderson, the British public was antagonistic toward Cook because he had been oversold by advance publicity. According to June, the single-named English actress who was co-starred with Joe, the trouble was that he failed both to adjust his material to his new audience and to adjust his voice to the huge theatre in which he appeared. Whatever the reason, for the only time in his life Joe was booed off the stage when he took his bows.

Even New York proved to be no longer as receptive as it had once been. The following year, Cook returned to Broadway in *Hold Your Horses,* a book musical that offered the comedian in the role of a hansom-cab driver in turn-of-the-century New York who becomes mayor by making incoherent speeches for the rival candidate. Though there was much joy in Joe's return to Broadway, the feeling was that even circus skills, gadgets, rambling anecdotes, and an ear-to-ear grin were insufficient to make up for the creaky plot and the dull production. The show lasted less than three months.

There was no denying, however, that Joe's hansom cab was one of his most brilliant—and practical—creations. Preceding the two-man horse, Magnolia, was a cow catcher to clean the street and behind a movable tail to swish flies. Magnolia even had a heart of gold which could be observed through an opening in its chest. To prepare the animal for the ride, Dave Chasen shoveled coal into the horse and oiled its bearings. Cabby Joe made himself comfortable above his passengers by outfitting his driving area with a hatrack, an automatic shoeshiner, a wash basin, a stove, a frying pan with an egg for his lunch, an alarm clock, a hatchet to cut bread, and even an elevator to transport him up to and down from his perch. That was only the beginning. He squirted seltzer water into a highball glass and made it foam like a geyser. He served five different kinds of drinks from the same cocktail shaker. Then, when a passenger requested a pousse-café, Joe rattled the same shaker and poured out a kitten. And after he was finished serving drinks and food, he pressed a button that caused all the china and glassware to disappear inside a table.

As usual, Dave Chasen thoroughly enjoyed being made the butt of Joe's humor. The new version of the Fuller Construction Company machine served this time to deposit the grinning stooge into a tank of water. And when Joe asked him to drop down to the corner to get a box of cigars, Chasen took the request literally and dropped through a trapdoor in the stage.

The disappointing run of *Hold Your Horses* made it apparent that Joe's style of musical comedy had become outdated. The theatre of the 1930s demanded more than innocent merriment, and Joe's familiar gags and technique were simply insufficient to carry a full evening's entertainment. The comedian tried adjusting his style for his last two stage appearances, but neither satisfied his old fans or brought him new ones.

When Joe Cook returned to Broadway in 1939, he no longer brought with him elaborate devices for achieving inane effects or Dave Chasen to bear the brunt of his jokes. He didn't even bring along songs and dancing girls. *Off to Buffalo*, in fact, was not even a Joe Cook vehicle, though he did receive star billing. It was a low-budget comedy (by Max Liebman and Allen Boretz) about a mousy secretary of a fraternal lodge in Brooklyn and the problems he has when he gets the bright idea of rounding up all his favorite vaudeville acts to put on a show for his fellow lodge members. Joe's part was that of the leader of the vaudeville troupe, with the lodge secretary played by Hume Cronyn in his first major role. Many observers found his antics as funny as Cook's.

But the play did give Joe the chance to perform his specialties when the entertainers audition their acts for Cronyn's committee. He juggled Indian clubs, told a rambling story about a man who ate turtle soup in Evansville, and he spun balls around the rim of an umbrella. Unfortunately, the show needed work, the producers didn't have enough money to delay the opening, and it folded in a week.

Rosamond Gilder, reviewing *Off to Buffalo* in *Theatre Arts* magazine, offered this appreciation of Joe Cook: "He is deft, agile, intimate. He draws his audience to him instead of hammering at it. His rounded eyebrows, his broad-gauge smile, his nimble wit excuse his rogueries. He is full of the milk of human kindness and will forgive those he has mulcted with so much charm that they are ready, even eager, to be defrauded again. 'I'll tell you all about it,' he says in answer to an embarassing question, and proceeds to recount a long, complex, factually and utterly irrelevant tale with such intensity, creating with voice, tempo and emphasis such interest and suspense, that everyone is enthralled. By the time he has reached his meaningless climax the force of his hearer's original anger is broken."

The following year Joe again tried a new approach. He came back on ice. One of the few feats of physical agility that the comedian had not mastered was ice skating, but that didn't stop him from being the stellar attraction of

Sonja Henie's *It Happens on Ice* when it played the Center Theatre. Though Cook's skating was kept to a minimum, he could be seen between the elaborate ice routines proffering the kind of stunts and inventions that had made him famous. There was no question that he was hampered by the cavernous theatre or the fact that ice skating fans were not necessarily Joe Cook fans. "What, I ask you," asked the *Journal-American*'s John Anderson, speaking for the critical majority, "can Joe Cook do in a refrigerator except freeze up? He ought to have steam heat and a warm heart, and good dry land."

He never did have them again, at least not in the theatre. Somehow the nimbleness and the timing were not quite what they should have been in *It Happens on Ice.* When he began missing clubs that he had once caught with ease, audiences, thinking it was part of the act, gave him some of his heartiest laughs. After discovering that he was suffering from Parkinson's disease, a progressive and incurable illness affecting the nervous system, Joe left the cast of the show (his final stage performance was on March 8, 1941), then bowed out as a performer as gracefully as he could with a simple ad in *Variety:* "Having been on the sick list for quite a while now, I have decided to quit the theatre." He sold his huge, rambling house near Lake Hopatcong, New Jersey (a celebrated mansion equipped with devices for practical jokes he enjoyed playing on guests), and moved to smaller quarters in Clinton, New York. There he lived in painful retirement for seventeen years. He died May 15, 1959, at the age of sixty-nine, without ever having imitated those four Hawaiians.

Jimmy Durante

OF ALL THE MAJOR CLOWNS who cavorted on Broadway, Jimmy Durante was the most frenzied, the most uncontrollable, and—to use a Durante-coined word—the most exibulant. He didn't simply appear on a stage he erupted on it. He was the nearest thing to a human volcano, a force of nature that could not be contained either by the part he was playing or by anyone else in his vicinity. Durante never seemed to try to be funny; his humor was nothing less than the natural, normal expression of an unquenchably comic spirit. When he told a gag "(I got a million of 'em, a million of 'em"), it wasn't the gag that got the laughs—or even the delivery—as much as it was Jimmy's sincere belief that the joke was not only funny but hilariously funny. And audiences howled because of their sincere belief in Jimmy.

That a performer of such manic intensity could also be so endearing was one of Jimmy's special gifts. Bert Lahr had much of his tireless, knockabout energy, but his explosiveness was never counterbalanced by such downright lovability and radiant good humor. Jimmy's innate sweetness, it should be noted, was apparent even in his wildest moments which usually found him railing against real or imagined adversaries that beset him on all sides. Jimmy was the grotesque Punchinello, mocking the pretensions of his betters, assaulting those who would assault him, loudly proclaiming his frustrations, yet at all times retaining that inherent winsomeness that made him as much a figure of affection as he was a figure of fun.

Jimmy's physical trademark—like W. C. Fields's—was his nose. Lending itself to self-caricature, it also lent itself to a nickname—variously "Schnoz," "Schnozzle," or "Schnozzola"—which became almost a part of his real name. Despite the early hurt of jeering classmates, Jimmy soon realized that his fleshy, somewhat extended nose could become very much a part of his comic personality, and he joked about it and sang about it. People didn't laugh simply because Jimmy had a long snout; they laughed because of Jimmy's attitude. Though it was his most prominent and most closely identified physical characteristic, Jimmy had others almost as distinctive. Primarily these were the voice and the walk. The voice was a croak, a rasp, a scratchy gargle, and his speech pattern was full of mangled pronunciations and the grammar of the streets. This too was an aspect of the comedian's personality since even off stage he had a hard time pronouncing words of more than three syllables. As for the walk, it was a slightly stooped strut with a jerky swaying motion accentuated by his wildly swinging arms. In height, Durante stood 5′7″, though he appeared shorter. His eyes were small and closely set, and his head was topped by a few curly, unruly strands of hair. At times he would break up his madcap activity by suddenly reacting to something he had said by coyly lowering his eyelids and pursing his lips in reaction to his own words. Most of the time his on-stage costume was a baggy suit and a battered fedora. He used the hat to punctuate his jokes by slamming it on the stage, and also to ward off opposing forces by throwing it at them.

In his early years as a night-club pianist and entertainer, Jimmy began writing stream-of-conscious lyrics to simple tunes using catch phrases that became his own untransferable property: "Ya know darn well I can do without Broadway, but can Broadway do without me?" . . . "Jimmy da Well-Dressed Man" . . . "Again ya turn-a" . . . "Who will be wid ya when I'm far away?" . . . "I ups ta him an' he ups ta me" . . . "Didja ever have da feelin' dat ya wanted ta go an' still had da feelin dat ya wanted ta stay?" And there were the meaningless exclamations—"Hot-cha-cha," "Inka-dinka-doo," "Umbriago"—that were as inseparable from the man as his nose. Jimmy also had a collection of trademark lines to reveal his exasperations which, no matter how often they were used, were enough to provoke roars of laughter from Durante devotees—"Ev'rybody wants ta get into de act" (if a waiter or chorus girl crosses his path during a number); "I'm mortified!"; "Surrounded by assassins!"; and "Stop da music!"

The son of a barber who had immigrated to New York from Salerno, James Francis Durante was born on the Lower East Side on February 10, 1893. As a boy he helped his father in his shop; later, when he had a newspaper route, he first became aware of the sounds emanating from the saloons along 14th Street. ("I peeps under da swingin' doors an' keeps t'inkin' dat da swellest job in da world belongs to da guy who bangs away at da piano. I wants to be him.") To help fulfil his son's ambition, Papa Durante bought a piano and hired a teacher. Jimmy didn't get along well with the teacher, but he took to the piano better than he did to academic studies. After the eighth grade he left school to become, among the other callings, a dishwasher and a photo-engraver. At night he earned extra money playing piano at local functions, and by the time he was seventeen he was a full-fledged piano pounder in Coney Island making

$25 per week. Now known as "Ragtime Jimmy," he soon found employment in night clubs in Manhattan.

At twenty-three, Jimmy led a five-piece jazz band at the Club Alamo in Harlem, where he hired—and eventually married—a singer named Jeanne Olson. There, too, he first teamed with a top-hatted, cake-walking song belter named Eddie Jackson. Jimmy was then still primarily a piano player; his comic gifts did not surface until he was joined by a third partner, a wiry vaudeville tap dancer named Lou Clayton. Jimmy by then had been talked into opening his own night club, the Club Durant, located over a garage on West 58th Street east of Broadway. (The misspelling in the club's name came about, according to the source one reads, either because Jimmy didn't have enough money to pay the sign painter for the final "E" or because the sign painter, a friend who donated his work free of charge, didn't know the correct spelling.)

It was Clayton's idea that the trio emphasize clowning rather than songs and dances, and it was also his idea that Jimmy be the chief clown, even though the act was billed as Clayton, Jackson and Durante. Club Durant flourished for six months until it was closed by Prohibition officers. By then the team's roughhouse act—revolving around Jimmy's gags, Jimmy's comic songs, Jimmy's nose, and Jimmy's frustrations—had become so popular that the boys were earning $3,000 a week in night clubs and vaudeville houses. They traded wisecracks, tossed remarks at friends in the audience, harmonized, danced a bit, tore a piano apart, and never let up for a second.

Lou, who was also the manager of the team, was anxious to expand its horizons. This, of course, meant Broadway. In the spring of 1927 he signed a contract with producer Charles B. Dillingham for the act to appear in a musical called *Ripples*. But *Ripples* was taking an extremely long time for its author to write, and the following year the trio accepted (but later returned) a $10,000 settlement to dissolve the contract. (*Ripples* was eventually—and briefly—presented early in 1930 with the leading role rewritten to fit acrobatic comedian Fred Stone.) In the fall of 1927, Clayton, Jackson and Durante were smashing pianos and house records at Loew's State Theatre; a year later they did the same thing at the Palace.

The manic brand of humor at which the boys excelled held no particular attraction for Florenz Ziegfeld, but the producer was well aware of their drawing power. At the urging of librettist-director William Anthony McGuire (he had been the author of *Ripples*), Ziegfeld signed the trio to appear in a backstage musical called *Show Girl*, which opened at the Ziegfeld Theatre on July 2, 1929. Ruby Keeler, who was billed as Ruby Keeler Jolson, having just married the "Mammy" singer, played an ambitious actress who becomes a *Follies* attraction; Jimmy was the property man with Clayton and Jackson as his assistants.

Eddie Jackson, Lou Clayton, Ruby Keeler, and Jimmy Durante in a scene from Durante's first Broadway musical, *Show Girl* (1929).*

Jimmy Durante and Lupe Velez in their hip-bumping sketch in *Strike Me Pink*.

The sheet music cover of Jimmy Durante's 1933 revue, *Strike Me Pink*.

Jimmy Durante and friend in the 1935 circus musical, *Jumbo*.*

Red, Hot and Blue!, which opened in 1936, had a stellar trio in Jimmy Durante, Ethel Merman, and Bob Hope. *

Though George Gershwin was the composer and Ira Gershwin and Gus Kahn were the co-lyricists of the score,* all of the Clayton, Jackson and Durante material consisted of songs that Jimmy had written for their night-club and vaudeville act. The show, in fact, was so tailored that it conveniently included a scene in a night club where the trio—called Gypsy, Deacon and Schnozzle—has come to audition. Following the introduction, Lou tap-dances down a flight of stairs at the left, Eddie struts down a flight of stairs at the right, and Jimmy, wearing a raccoon coat, bustled through a center door and dashes to the footlights. After his partners have pounded him on the back in an excessive show of camaraderie, Jimmy confesses he's not happy. In fact he's mortified. "Whaddaya t'ink happened?" he asks, shaking his head. "I drove up in front of a café in dat new Ford of mine and I'm coverin' da engine wid a blanket so da sun won't shine on da carburator, an' a little kid lookin' on says, 'I seen what it was, mister. Ya didn't have ta hide it.' He was a cute kid so I kicked him right in da head." Lou grabs Jimmy's nose, Jimmy rushes over to the piano, and Lou knocks his hat off. As Jimmy plays, Eddie prances about bellowing a song called "Because They All Love You." After Lou does his soft-shoe specialty, Jimmy stands at the piano and plays a few bars with his right hand. Lou knocks his hat off again, and Jimmy puts on a derby. "Say, Lou," Jimmy asks, "remember da trick we used ta do with da hat?" "Yeah." "Well, I can't do it no more."

At the piano, Eddie joins Jimmy in singing "Who Will Be with You When I'm Far Away," followed by this exchange:

EDDIE: Where are ya goin' tomorra?

JIMMY: I'm goin' to an insane asylum.

EDDIE: An insane asylum?

JIMMY: Yeah, I'm gonna get me a ravin' beauty!

The final dialogue in the routine concerns what happened when Jimmy asked a department-store floorwalker to mind his little dog:

JIMMY: And can you imagine, when I came back I actually saw him kickin' my dog.

EDDIE: The big brute.

JIMMY: So I ups to him.

* Durante was scheduled to sing the show's standout number, "Liza," but for whatever reason by the time the show opened out of town, it was given to Nick Lucas. Actually, newlywed Al Jolson on more than one occasion stood up from his seat in the audience and sang "Liza" to his bride as she tapped her way down a flight of stairs. Durante, however, did get to sing it for a week during the Broadway run.

EDDIE: Easy, Schnozzle.

JIMMY: I says, "What's the idea of kickin' my little dog? He wouldn't bite you."

EDDIE: What did he say?

JIMMY: He said, "I don't know about him bitin' me, but he had his leg up ready to kick me."

Durante throws his hat and cane on the floor as the trio stride off, arms waving madly, and ear-to-ear grins on their faces.

Earlier in *Show Girl*, in a backstage scene, Jimmy and his cohorts sang "Can Broadway Do Without Me?"—a rambling account of how our Jimmy was called on to replace two ailing headliners, Eddie Leonard and Al Jolson, and how, just by stamping his foot, he had become the inspiration for "The Varsity Drag" in *Good News*. In fact, according to Durante, the step was so closely identified with him that when word got around the "tee-ater" that he was there ("It spread like wildflower"), the audience rose "and shouted to da performers, 'Impostigators! We want Jimmy da Well-Dressed Man!' "

In Act II, with equal lack of motivation, Durante sailed into "I Ups to Him," his tale of an encounter with a man who bumped into him on Broadway and pushed him off the sidewalk. To avoid a brawl, Jimmy beat a hasty retreat and stopped off at Liggett's drug store for a yeast cake. There he came up against a tough Westerner who roughed up Jimmy just to see if he could take it. Years later, while lying on the beach at Waikiki, Jimmy was again accosted by his first assailant, who continued to taunt him. "So I ups to him," sang Jimmy, "an' he ups to me. So I goes my way, an' he goes da way of all flesh."

It was also in *Show Girl* that Jimmy recited his celebrated tribute to a small domicile, "I Like a One-Room Home." The scene was the street in front of a row of Brooklyn tenements where Frank McHugh, as Ruby Keeler's suitor, has just moved. Frank tells Jimmy he likes a big house, but Jimmy has other ideas:

"I like a one-room home. You might like a two-room, t'ree-room, four-room, five-room, six-room, seven-room, eight-room home. I only have a one-room home. An' if you was to ask me why I have not got a two-room, t'ree-room, four-room, five-room, six-room, seven-room, eight-room home, I would answer you because I do not want a two-room, t'ree-room, four-room, five-room, six-room, seven-room, eight-room home. I only got a one-room home. You like your two-room, t'ree-room, four-room, five-room, six-room, seven-room, eight-room home. Keep your two-room, t'ree-room, four-room, five-room, six-room, seven-room, eight-room home, an' let me have my one-room home. Because da furniture I have does not fit in a two-room, t'ree-room, four-room, five-room, six-room, seven-room, eight-room home."

Jimmy's intensely serious manner and his deliberate, rhythmic repetition of words, invariably brought down the house. It also brought charges of copyright infringement. Durante maintained that he got the idea for the "one-room home" bit from an inebriated friend babbling about the impecunious though happy early years of his marriage. A poet named Alfred Kreymborg, however, insisted that it was stolen from his poem, "I Got a One-Room House," and later sued Durante and NBC when Jimmy performed it on the radio. The judge ruled in Durante's favor, but NBC agreed to an out-of-court settlement with Kreymborg.

Even though *Show Girl* barely managed a three-month run, the team of Clayton, Jackson and Durante had made an almost unqualified hit on Broadway. As Finley Peter Dunne wrote in the *World:* "Their unquenchable energy, their shouted quips, their manifold comic songs (the same as always) run through the show from start to finish. Keeping it moving at high speed, Durante and his confreres are given plenty of work to do, but never a whit too much." Somewhat less impressed was the *Times*'s Brooks Atkinson: "Their transformation into *Show Girl* is not altogether sublime. Durante's sizzling energy can galvanize any audience, but his spluttering, insane material does not melt gracefully into a musical-comedy book. His personality, however, batters its way through all barriers."

Jimmy's second musical (after making his first film, *Roadhouse Nights*) was *The New Yorkers.* Again he found himself in a show involving some major talents: Cole Porter for the songs (including "Love for Sale" and "I Happen To Like New York"), Herbert Fields for the libretto, and Hope Williams, Frances Williams, Charles King, Ann Pennington, Marie Cahill, Richard Carle, and Fred Waring's Pennsylvanians in the cast. The work itself was rather daring in that it was concerned with a group of totally amoral characters representing both high and low society. There were stops at a bootleg factory, a gangster-owned night club, a street in Harlem, Reuben's Restaurant on Madison Avenue and 59th Street, an apartment house on Park Avenue, Sing Sing Prison (where Jimmy teaches classes in safecracking, forgery, and pickpocketing), and—following a prison break—a wedding ceremony in Miami (where the maid of honor carries a bouquet made of pistols and daggers, and the bridesmaids carry small bombs).

As he had done for the team's assignments in *Show Girl,* Jimmy wrote all the songs for Clayton, Jackson and Durante (here known as Monahan, Gregory and Deegan). And once again they were in a Broadway show that provided the trio with a night-club sequence in which to do their stuff. In their first routine at the Club Toro, the boys introduced a dance step, "The Hot Patata" ("Now first ya tip your hat-a, Then ya give it this-a, Now ya put on your hat-a, Then ya give it that-a"), which contained Durante's insistent refrain, "Again ya turn-a." There was also the lunatic number in the prison classroom in which Jimmy discoursed on his worldwide pursuit of scientific research called "Data" ("Da next day—breakfast with Einstein, chatted good-naturedly about da fourth dimension"). The search soon found Durante, along with Clayton and Jackson, as passengers on a ship during a raging storm. To simulate the frantic occasion, his partners squirted Jimmy with water, the trio made a mad dash for a lifeboat, and the scene ended with the men rowing furiously for the horizon.

Jimmy's outstanding number was unquestionably the one called "Wood," originally performed a year earlier at the Palace and re-created here as the first-act finale. Durante had composed it to the tempo of Kipling's poem, "Boots," after reading a magazine ad for a lumber company extolling the importance of wood in the building of America. The routine was introduced in *The New Yorkers* in a scene in which the entire company, led by Clayton and Jackson, accuses Durante of having a head made of wood. Suddenly the downcast Jimmy jerks his head high, smiles a proud smile, and tells them all that they have just paid him a compliment. Why? Because "without wood there would be no America," insists Jimmy. He then enumerates all the products made of wood, such as the ships that brought the settlers across the stormy Atlantic, the sturdy log cabins ("which were gigantic"), schoolhouses, town halls, bridges, warehouses, churches, covered wagons, and railroad ties. Then becoming truly rhapsodic about his subject, Jimmy wins over his accusers with this final tribute:

> Wood is warm an' alive to da touch—da handle of a tool, da arm of a chair.
> Wood is stronger pound for pound than any other material, an' it can be carved like a turkey.
> Wood is endearin' and da supply is unlimited.
> An' when one talks about da unlimited
> He's talkin' about somet'in' he knows not what.

Once he concludes the panegyric, Jimmy leads his two buddies and the rest of the cast into transforming the stage into a dumping ground for all kinds of wood products—furniture, boxes, doors, pushcarts, barrels, canoes, barber poles, rickshaws, even an outhouse, plus, from the orchestra pit, violins, a bass drum, and parts of a piano. All the while everyone is feverishly chanting, "Wood . . . wood . . . wood . . . wood . . ." until there is barely enough room for anyone to move. According to Brooks Atkinson, "During that scene, *The New Yorkers* is as overpoweringly funny as a weak-muscled theatregoer can endure," and Gene Fowler, Durante's biographer, unequivocally called it the greatest act the trio ever did.*

Jimmy Durante's success in *The New Yorkers* led to a five-year contract with M-G-M that did not include Clayton or Jackson. At first Jimmy refused the terms, but the economic realities of the Depression years forced him to accept (though he always split his salary three ways), with Clayton going along as his manager and Jackson as his secretary. Between 1932 and 1941, Durante made twenty-one movies, few of which took full advantage of his special gifts.

In 1933, about two years after his final appearance in *The New Yorkers*, Jimmy returned to Broadway in the Ray Henderson–Lew Brown revue *Strike Me Pink*. Originally, the show had had a different concept, and Durante was not in the cast. Under the title *Foward March*, it had been conceived as a showcase for unknown talent, but its prospects looked grim after a poorly received tryout in Pittsburgh. Mobster Waxey Gordon, the revue's sole financial backer, insisted that only the infusion of star talent could save the show, and it was retitled, rewritten, restaged, and recast with Jimmy Durante, Lupe Velez, and Hope Williams in the leads. By the time it opened in New York in March, *Strike Me Pink* had already cost Gordon $150,000.

Jimmy appeared in the musical minus his partners or any of his familiar night-club material, but he dominated the show right from the beginning. In the first scene the featured players were introduced as if they were being interviewed at a film premiere by a bumbling radio announcer (Roy Atwell). Jimmy was saved for last. In fact, when the audience first became aware of him, he was in the left aisle raging at an usher who was trying to prevent him from joining the cast on stage. What caused him particular distress was that Eddie Garr, a Jimmy Durante impersonator, was being introduced as the real Jimmy Durante. Tearing himself loose from the usher, the comedian dashed down the aisle and up onto the stage yelling, "Wait a minute! Wait a minute! Dat ain't Jimmy. I'm Jimmy da Well-Dressed Man!" Though Durante and Garr each insisted that he was the genuine article, Garr soon admitted to being the understudy, and Jimmy acknowledged the audience's applause by expressing his pleasure at being back on Broadway.

Jimmy also took advantage of the theatre's aisles during at least one performance of a sketch with Lupe Velez. It also showed how Jimmy's violent humor could occasionally go too far. In the skit, Miss Velez played a temperamental actress dissatisfied with the chair she had been given to sit on, and Durante was the show's property man ("Oh, da temperament of dese people! Me what goes out wid de elite! De intelligentsia!"). The scene called for Lupe

*A later show-stopping number extolling the uses of wood, "Song of the Woodman," was written by Harold Arlen and E. Y. Harburg for Bert Lahr in *The Show Is On* in 1936.

and Jimmy to bump hips and one night Jimmy put a monkey wrench in his overalls pocket. Lupe screamed with pain, reached into Jimmy's pocket, grabbed the wrench, and proceeded to chase Jimmy up and down the aisles. When she finally caught him, the two wrestled briefly on the stage until Durante ended the brawl by tossing Miss Velez into the orchestra pit.

Though reviews were mixed regarding the show, all the critics except the *News's* Burns Mantle ("He is better in pictures than he is in person") gave Jimmy an enthusiastic welcome. In the words of John Mason Brown in the *Post:* "Jimmy Durante is the most energetic of our comedians. There is no tiring him. From the eyebrows above his far-famed schnozzle to the heels of his desperately stamping feet, he is a powerhouse in mufti. His pallid face does not know what relaxation means. Neither does his wiry body. Every minute of the time that he is on the stage he is frantically at work, jamming hats down over his disordered hair, pursuing jokes with the relentless vengeance of a bloodhound, shaking his head and rolling his eyes until they have achieved that smiling ecstasy which is required for one of his renowned 'Hot-cha-chas.' " *Strike Me Pink* could have had a longer run than its 105 performances but Jimmy's film commitments cut short the engagement.

Jumbo, in which Durante appeared in 1935, was one of the most ambitious musicals ever mounted in New York. Showman Billy Rose's concept was to combine the greatest international circus attractions—aerialists, acrobats, clowns, equestrians, animal trainers—with a musical-comedy book show. Since no ordinary theatre would do, Rose took over the mammoth Hippodrome, which he had thoroughly rebuilt with arena-type seats and one circus ring that also served as the playing area. For his libretto he secured Charles MacArthur and Ben Hecht, and for his score he had Richard Rodgers and Lorenz Hart (whose contributions included "Little Girl Blue," "The Most Beautiful Girl in the World," and "My Romance"). John Murray Anderson and George Abbott (it was Abbott's first musical) were in charge of putting the whole thing together. Originally scheduled to open in April, the official premiere was postponed so often that the producer even signed an affidavit swearing that Jumbo would indeed open on November 15.

Durante's role was that of Claudius "Brainy" Bowers, a press agent for a floundering circus, who gets into trouble by talking too much, lights matches to help him concentrate, instructs the show's romantic leads on the proper way to sing "My Romance" ("Not dem piccolo high notes but da cry of an eagle"), and even commits arson so that the circus owner can collect the insurance money.

One of the scenes in *Jumbo* has become a comedy classic. Because the circus is in debt for back taxes, a federal marshal has taken it over to sell its assets at auction. "Brainy" Bowers, hoping to save at least the company's prize elephant, the eponymous Jumbo, is tiptoeing out of the fair grounds with the pachyderm on a leash when the marshal suddenly appears. "Hey!" he bellows, "Where are ya goin' with that elephant?" Feeling that silence will somehow render the animal invisible, Brainy cautions his charge, "Shhh . . . Not a word, not a word," and keeps on moving. When the marshal again demands to know where he's going, Brainy stops, adopts an expression of innocent incomprehension, takes a quick look around, and asks, "What elephant?"

But the scene does not end there. In order to prove that Jumbo is his own personal pet and does not belong to the circus, Brainy stretches on the floor and has the elephant step over him so that its right forefoot comes within an inch of his face.

Though credited to Rodgers and Hart, Jimmy's songs in *Jumbo* came as close as any numbers of his own creation to capturing the rambling, free-form spirit of Durante's night-club routines. In one of them, titled "Laugh," the comedian gave this advice to the downcast heroine:

Ya gotta laugh, laugh, laugh,
When t'ings are catastrophic.
Ya gotta laugh, laugh, laugh,
An' take it philostrophic.

Jimmy went on to recall—in a song recalling his own "I Ups to Him"—the day he laughed a guy out of handing him a "suppeenee," and the time in the lingerie section of a department store when a guy accused him of being "a chemise fancier, a twiddle twa." Though that encounter ended in a fight, Jimmy still could offer the philostrophic advice to "be like a Pagliacci an' laugh."

Another Durante song, "Women," resulted from a brainstorm in which the excitable press agent comes to the conclusion that what the circus most needs to attract customers is "a galaxity of femininity, hand-picked from da rose gardens of Broadway." His unassailable argument:

Put twelve acrobats up dere, hangin' by their toes,
Fill da ring wid elephants, lions, tigers, kangaroos.
I'll even let ya t'row in a penguin.
Then have a beautiful voluptuous woman walk down da center aisle
of da ring. *(On cue, a girl walks down.)*
As illustrated.
What do they look at? It don't need an answer.
Do what I tell ya and ya'll have da greatest show dis side of oblivion.

As the song ends, the stage is filled with the prescribed "galaxity" of dancers, big-top specialists, and show girls who are borne aloft in large metal hoops in which they perform their somewhat constricted maneuvers. As one of the aerialists, named Barbette, goes through a variety of daring stunts, "Brainy" Bowers proudly announces, "Da woman supreme! Nature's masterpiece! A combination of genius an' continuity! A moonbeam in de air an' a Garbo on da ground!" With that Barbette removes his wig to reveal that he's a he, and Brainy, totally flabbergasted, can only mutter, "Betrayed!"

Red, Hot and Blue!, which opened in New York just six months after *Jumbo* had closed, was the first of two book musicals in which Jimmy Durante was co-starred with Ethel Merman. Cross-starred would be more accurate since Merman and Durante—or their agents—insisted on the top billing position (the left side above the show's title) and a compromise was worked out with their names crossed. Actually, it really wasn't much of a compromise since the positioning still favored Durante:

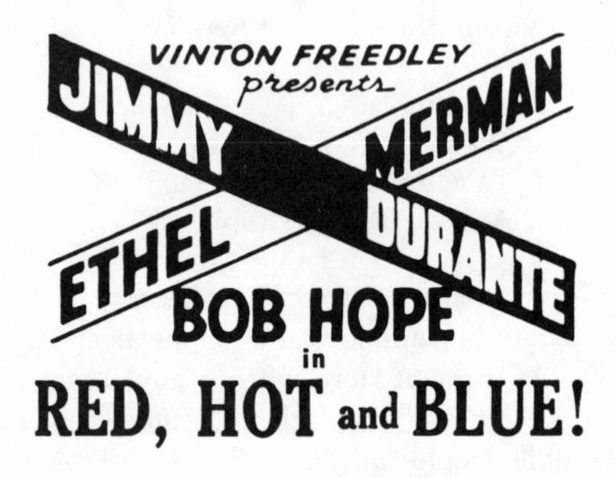

Producer Vinton Freedley's original idea was based on what must have struck him as a surefire formula: take the stars and writers of his hit, *Anything Goes*—Merman, William Gaxton, Victor Moore, Cole Porter, Howard Lindsay, and Russel Crouse—and involve them in a political satire suggesting the classic Gaxton-Moore spoof, *Of Thee I Sing.* But after Gaxton overheard librettists Lindsay and Crouse convincing Miss Merman that her role would be the most important, he and Moore withdrew and were replaced by Bob Hope and Jimmy Durante (after the producer had failed to get Willie Howard for the Moore-intended role). The revised script was a simple-minded but complicated tale in which Merman played "Nails" O'Reilly Duquesne, a former manicurist now a wealthy widow, who gets permission for Durante, as Bertram "Policy" Pinkle, an inmate of Larks Nest Prison, to help her run a national lottery for charity. The idea of the lottery, devised by lawyer Bob Hope, involves revealing the identity of a young lady who had sat on a waffle iron when she was four. Jimmy, however, is miserable away from the luxurious surroundings of dear old Larks

Nest, where he is the captain of the polo team, and does everything he can to be recommitted.

The script, as might be expected, gave Jimmy a number of gags that capitalized on his untutored image. Such as:

DURANTE: Ya know what Aristotle said about da legal mind?
MERMAN: No.
DURANTE: Dat makes us even.

And

DURANTE: Don't cry. Please don't cry. Don't be lugubrious.
MERMAN: What does lugubrious mean?
DURANTE: Go ahead an' cry.

Jimmy also praised his father as a great idea man "an' I always try to follow in his fingerprints." Later, when costumed as George Washington (don't ask why) for a White House wedding, he referred to himself as "an old Valley Forger."

His main comic opportunity came during the scene of the Senate Committee investigating the lottery, in which he both testified and served as his own lawyer. As he relentlessly cross-examined himself, Jimmy dashed back and forth to the witness chair after each question, creating a flurry of comic movement that even had him tripping over his own feet. The easily frazzled witness confessed his guilt, then complimented the lawyer ("How can I hide anyt'in' from such a mastermind?"), and pleaded to be sent back to jail.

Durante had one song, "A Little Skipper from Heaven Above," that unquestionably had to be seen to be appreciated. In a totally unmotivated sequence, Jimmy stops reporters' questions by announcing "I can tell ya about a predicament dat was nothin' short of whimsical," then sings the rollicking tale of a sea captain who proclaims to his men, "I'm about to become a mother, I'm only a girl not a boy."

Jimmy's critical reception was more enthusiastic than ever before. "Almost whenever he appears," wrote John Mason Brown, "is Jimmy Durante as you have always dreamed of his being but as he has seldom had the chance to be. His is more than uproarious. He is convulsingly funny—so sidesplittingly funny that not seeing him would be an act of self-denial of which no wise playgoer would think of being guilty." And according to John Anderson in the *Journal*, "Mr. Durante is at his best when the audience is in an uproar. Maybe the noise drives him crazier than ever, and he reaches those pinnacles of gleam-eyed insanity in which he stands happily alone among the clowns of the theatre. The show itself never reaches such whirlwinds and Mr. Durante makes the ascent under his own steam."

When Jimmy was reunited with Ethel Merman in 1939 for *Stars in Your Eyes,* there was no hassle over billing as there had been for *Red, Hot and Blue!* Ethel was given—or took—the coveted left-side-above-title position. The musical itself was originally conceived by composer Arthur Schwartz, lyricist Dorothy Fields, and librettist J. P. McEvoy as something of a left-wing satire on Hollywood's extravagant style of movie-making. Called *Swing to the Left,* the show at first had Jimmy as a labor organizer in the studio where Ethel is the star attraction. By the time director Joshua Logan arrived on the scene, however, he managed to convince everyone that the social-significance slant was wrong for a star-studded musical comedy. What they changed it into was a fairly conventional work with Durante's role that of a studio idea man. Even the names by which Ethel Merman and Durante were known in the show—Jeanette Adair and Bill—lacked the proper comic flair.

Jimmy's role was also something of a departure from the past in that instead of being an explosive, dim-witted character, his part was now that of a highly esteemed trouble-shooter whose ideas—no matter how outlandish—are greeted by rapturous praise. All this, as he pridefully explained in the song, "Self-Made Man," came about because "I made me what I am today":

> What made Heifetz? A fiddle!
> What made Ben Bernie? A cigar!
> What made Mr. Moses? A parkway!
> What made history? Hedy Lamarr!
> But as for me . . .
> No one put any eggs in my basket,
> No one gave me a tisket . . . or a tasket.

It was in *Stars in Your Eyes* that Jimmy first showed his ability at transforming songs of a generally sophisticated show-tune style into his own special property. In the first, "Terribly Attractive," he was paired with Mildred Natwick as a sharp-tongued Dorothy Parker-type screenwriter. After singing each other's praises on a deserted sound stage, they perform an uninhibited, show-stopping rumba and tango. A two-seater motor car is then rolled onto the set and Jimmy gallantly offers to take Mildred for an imaginary spin. As they get into the car, a screen descends behind them showing the kind of process shots that always appear when, presumably, an automobile is in motion during a movie. Beginning placidly enough on Hollywood Boulevard, the scenes abruptly change to a country road with a motorcycle policeman chasing them, a New York street beneath an elevated railway, a tunnel, the Western plains with a dozen cowboys shooting from horses, a charging locomotive, a flotilla of battleships, and, finally, a trolley car which barely avoids hitting them. All the while Durante and Natwick are responding with increasing terror until Miss Natwick manages to climb out of the car (still parked, of course), with her hat fallen over her nose, her knees buckling, and one shoe off. Dazedly, she looks around, holds up one limp finger, and staggers off the stage weakly calling, "Taxi! Taxi!"

In the second act, Richard Carlson, playing the role of an idealistic screenwriter, has fallen asleep and dreams that he is not only in charge of production at Monotone Pictures but that he is also acting the part of an unhappy czar in a Russian epic. Ethel appears in a long ermine-trimmed gown with a large jewel-encrusted headdress, and Jimmy comes out wearing a silk shirt, baggy peasant pants, black boots, and a fur hat. They have come to cheer up the moody monarch with the upbeat sentiments of "It's All Yours" ("Just leave your cloister, baby, stop mopin',/ The world's your oyster ready to open"), with Ethel doing the singing and Jimmy tearing apart a piano. During a reprise, Jimmy interpolates spoken cheer-up bromides after almost every line. As the orchestra plays the chorus a third time, he tries making the czar laugh by telling a story: "I went to see my dentist de other day, an' a tough guy gets out of da chair. So I says to him, 'Let me see your teeth.' Dis side where da dentist was workin' *[pointing to the left side]* was perfect. Dis side *[pointing to the right side]* dere was one tooth missin', half a tooth missin', one tooth, half a tooth. *[Now the rhythm is emphasized as Ethel swings her hips to the beating of a drum.]* What goes on? So I says, 'If dis side is perfect, why don't ya get da dentist to fix da other side?' He says, 'Why should I? Dis is da side I talk out of.' " With that Ethel and Jimmy double over with laughter, slapping their legs and each other. After another gag the telephone rings, and Jimmy answers it. "Hello, is dis da meat market? Well, meet my wife at four o'clock." Jimmy throws the phone over his shoulder, and he and Ethel truck off the stage singing the last bars of the song.

The audience, of course, sensing that there are still more quips to come, gives them an ovation. Back come Ethel and Jimmy, strutting to the song's rhythm pounded out by the drummer. Jimmy almost tumbles over the footlights before Ethel pulls him back.

JIMMY: I ran into Professor Einstein today, and how do ya t'ink I found him?

ETHEL: I don't know. How did you find him?

JIMMY: I just pushed back his hair—an' dere he was.

Both Ethel and Jimmy are now laughing almost uncontrollably. Following a few more sallies, they again strut off stage singing. A third wave of thunderous applause brings them back—only this time it's Ethel who almost falls over the footlights. Jimmy tells her, "I'm walkin' down da street today and I bumps into a guy. So I apologizes. But he ain't satisfied. He demands an autopsy. I sees an openin' *[With that there's a loud crash from the drummer.]* I'm flat on my back!" Jimmy takes off his hat and hurls it at the orchestra, and he and Ethel again break into gales of laughter. Ethel says, "I was riding on a bus today and I asked the conductor, 'Does this bus go over the Queensborough Bridge?' And he said, 'Lady, if it don't we're both gonna get a hell of a duckin'!" More hysterics, more back slapping, more singing—and exeunt to prolonged laughter, applause, cheers, and whistles.

For his last Broadway show, *Keep Off the Grass,* in 1940, Jimmy appeared in a revue with a theme—or at least a unified setting, since all the scenes took place in Central Park. Though the cast was well received (it included Ray Bolger, Ilka Chase, Jane Froman, Virginia O'Brien, Larry Adler, José Limon, and a rotund comic named Jackie Gleason), the show was considered weak in material (or, as Abel Green put it in *Variety,* "It's not a good $4.40 buy").

Incongruity, of course, is an important part of a comedian's technique. Durante as a polo-playing convict in *Red, Hot and Blue!* or as a studio idea man in *Stars in Your Eyes* evoked laughs because the roles were totally at odds with his image as an unlettered innocent. In *Keep Off the Grass,* almost everything he did gave him the opportunity to appear in a completely incongruous role. He was Dr. Kildare, a tree surgeon ("The Pasteur of the Pastures"); played Romeo to Ilka Chase's Juliet in a balcony scene supposedly directed by the slapstick team of Olsen and Johnson; impersonated President Roosevelt in a sketch in which the First Family somehow merged with the red-haired Clarence Day family of *Life with Father;* passed himself off as an erudite Clifton Fadiman in a take-off on radio's *Information, Please* with monkeys as the customary intellectual panel members; and appeared as a highly unlikely Clark Gable singing, with Miss Chase and Ray Bolger, about the making of *Gone with the Wind.*

Possibly Jimmy's most effective number was a musical variation on his oft-proclaimed line, "I'm Jimmy da Well-Dressed Man." In Jimmy McHugh and Howard Dietz's "A Fugitive from Esquire," Durante, in top hat, white tie, and tails, was seen leaning against a statue of Esky, the *Esquire* magazine symbol. After making and breaking dates with six formally clad show girls holding fur-covered telephones, he turned to the audience to recall two of his most humiliating experiences:

"Feelin' kinda dapper in my newest creation, I stepped out for a promenade. Swingin' my Malacca umbrella nonchalantly, I went strollin' down da boulevard. I know I was creatin' a sensation and given' da pedestrians a treat. As I stepped into an open Baruch, who do ya t'ink I meet? Cole Porter? No! Noël Coward? No! Cholly Knickerbocker? You're gettin' warmer. Lucius Beebe? Yes, sir! Lucius Beebe. I could visualize Lucius writin' me up in his column sayin': Quote. Glimpsed at Sutton and Beekman was Jimmy Durante wearin' a two-toned striped herringbone with a three-button effect. Bottom button at da bottom. On his notched lapel for a boutonniere was a red carnation, an' settin' off his ascot tie was his tan reversed-calf shoes. At last Adolphe Menjou has been scuttled. Unquote."

But to Jimmy's dismay, Beebe was unimpressed with his finery and told him he didn't belong. ("What a blow! Ignored, harassed and shunned by dose in da know.") Describing himself as a "broken, shattered fashion plate," Jimmy then recalled the time he was snubbed by the Harvard "Regretta." Both hurt and bewildered, Jimmy then had four valets go into the audience to display his sartorial elegance ("Why, I spend $30,000 just for mothballs"), and the scene ended with bundles of clothes descending on the stage from the wings and the flies.

After a disappointing run, *Keep Off the Grass* closed on June 29, 1940. Though Durante was to live almost forty years after that date, it marked his final appearance on the Broadway stage. The reason may have had something to do with the changing nature of the musical theatre, led by *Oklahoma!* in 1943, and the fact that Jimmy was possibly even more comfortable as a night-club clown than he was as a Broadway clown.

After going through a relatively lean period, made doubly painful by the prolonged illness and death of his wife, Jimmy re-emerged as a topflight entertainer in 1943 when he revived the Clayton, Jackson, and Durante act at the Copacabana night club in New York. This led to a radio series with Garry Moore and eventually his own television series. In the meantime, Hollywood again beckoned. This time the parts were better and Jimmy even got a chance to appear in MGM's 1962 screen version of *Jumbo* (along with Doris Day and Martha Raye). In his last years, though, he was primarily a night-club, television and recording attraction, with his video signature, "Goodnight, Mrs. Calabash, wherever you are," taking its place among the immortal Durante phrases. In retirement from 1972 because of a stroke, he died in Santa Monica on January 28, 1980, when he was almost eighty-seven.

Jimmy Durante was an unlettered mug and was happy to remain one. There is no indication that he ever brooded over his art or tried breaking out of his trademark routines. At heart, he was a night-club buffoon who reveled in the wild disorders of a madcap floor show in a smoke-filled room. When transplanted to Broadway in the seven musicals in which he appeared, his disheveled spirit was never bounded by a script or even a proscenium arch. To see Jimmy Durante on the stage was to see a totally unfettered free soul, the most natural and unself-conscious clown who ever trod the boards—and then tore them apart.

W.C. Fields

THERE SEEMS TO HAVE BEEN two general categories of male clowns on Broadway. In one category were those who inhabited a fantasy world of their own making. Their pranks, outlook, and bizarre makeup and costumes all proclaimed their indifference to everyday life and its problems. Diverse though they may have been, such clowns as Ed Wynn, Bobby Clark, Joe Cook, and the Marx Brothers lived in their own special universe, no matter the predicaments in which they found themselves. In the other category, clowns usually less outlandish in appearance and personality made the most of recognizable human situations to achieve a bond of empathy—or at least of recognition and understanding—between themselves and their audience. Buffoons such as Willie Howard, Victor Moore, Jimmy Durante, and Bert Lahr were comic Everymen, forever grappling with inanimate objects that confuse and break down, situations for which they were totally ill-equipped and unprepared, and all manner of real or imagined antagonists.

Of those comedians who had to deal with life's unending problems, surely W. C. Fields projected as expertly honed a characterization as any performer who ever extracted humor from adversity. Beset on all sides by a multitude of irritants and unable to cope with the demands on his time and patience, Fields projected a character continually at the mercy of unmanageable forces. Through it all, while he may have been smoldering with frustration, he never exploded with rage. Adopting a defensive pose of one who is high above the battle, he underplayed his reaction to misfortune by seeming to have his mind on other, more pressing matters as he vainly tried to hold on to whatever vestiges of self-respect he possessed.

While there may have been a good deal of slapstick and comic exaggeration in his early revue sketches, by the mid-1920s Fields had evolved the basic character by which he was identified, not only on stage but also on radio and in films. Primarily it was that of a 20th-century Falstaff, a blustering, jovial fraud with an orotund way with words and an obsequiously polite manner. Good fortune was to be obtained only through deception, and an insincere teeth-baring smile could never fail to impress gullible ladies. His disdain for lesser mortals was allowed expression only through the comments he muttered under his breath or through an imperious wave of his hand.

There was nothing sentimental or lovable about W. C. Fields. We knew that behind that pompous dignity was a thoroughly maculate character. Yet though Fields never courted our affection or our sympathy, we could understand his plight and the bogus nonchalance with which he tried to conceal his irritations. Curiously, his stage career found him in only two productions—*Poppy* and *Ballyhoo*—that allowed him the opportunity to portray a sustained role throughout a full-length musical play. The other twelve musicals in which he appeared (including two not seen on Broadway) followed a pattern of presenting Fields in specialty routines, initially as a comic juggler but mostly in brief revue sketches that he usually wrote himself.

Physically, Fields was a beefy man with sandy-colored hair, beady blue eyes, and a round face dominated by a bulbaceous nose (which became even more bulbaceous as the years went by). His voice, emanating slightly from the side of his mouth, had the quality of rusty taffy. During most of his career on stage—and occasionally even off it—he was addicted to wearing clip-on mustaches of varying shapes and sizes that he affixed in his nostrils much as one would attach a clip-on bowtie under one's collar. The image that he projected, that of a basically even-tempered person made sour and suspicious by countless battles with alien forces, was not entirely limited to the stage. It was, in fact, an extension of the man himself, brought on by a youth of poverty, fear, and loneliness.

There is some dispute regarding the comedian's birth date. Robert Louis Taylor, Fields's biographer, put it at April 9, 1879. Ronald J. Fields, W.C.'s grandson and compiler of *W. C. Fields by Himself*, mostly a collection of letters and scripts, maintained that "according to the best records" the date was January 29, 1880. There is no dispute that his real name was William Claude Dukinfield, or that his friends called him Bill, or that he grew up in the Germantown section of Philadelphia. Fields's father was a tough immigrant Cockney who sold vegetables and fruit from a street barrel; his mother was a buxom woman with a keen sense of humor, whose muttered asides about people and events later influenced her son's speaking style. As a boy, young Bill helped his father at work, but after a fight resulting in the lad crowning his parent with a wooden box he felt that a separation would be in his own best interest. Fields later wrote, "My earliest recollections are none too pleasant. I ran away from home when I was eleven. Well, I didn't run away exactly. I just didn't

W. C. Fields, Will Rogers, Eddie Cantor, and Harry Kelly show their admiration for Lillian Lorraine in the *Ziegfeld Follies of 1918*.

The "Joy Ride" sketch in the 1925 *Ziegfeld Follies*, with Martha Lorber, W. C. Fields, and Ray Dooley. *

NEW AMSTERDAM THEATRE

42D STREET
WEST OF BROADWAY

KLAW & ERLANGER
MANAGERS

THIS THEATRE, UNDER NORMAL CONDITIONS, WITH EVERY SEAT OCCUPIED, CAN BE EMPTIED IN LESS THAN THREE MINUTES. LOOK AROUND NOW, CHOOSE THE NEAREST EXIT TO YOUR SEAT, AND IN CASE OF DISTURBANCE OF ANY KIND, TO AVOID THE DANGERS OF PANIC, WALK (DO NOT RUN) TO THAT EXIT.

WEEK BEGINNING MONDAY EVENING, AUGUST 23, 1915.
Matinees Wednesday and Saturday.

ZIEGFELD FOLLIES
1915

Devised and produced under the personal direction of
F. ZIEGFELD, JR.
Staged by Julian Mitchell and Leon Errol.
Lines and Lyrics by Channing Pollock, Rennold Wolf and Gene Buck.
Music by Louis Hirsch and David Stamper.
Scenery painted by Joseph Urban, of Vienna.

ACT I.

Scene 1—**UNDER THE SEA.**

Note—Owing to the mechanical massiveness of this scene, it can only be exhibited advantageously at the beginning of performance—8.20 sharp. The stage setting is the greatest Mr. Ziegfeld has ever presented.

Channel Belle .... Kay Laurell
Submarine Pilot .... Chas. Purcell
Mermaids........................Misses Koerner, Hart, Touraine

Scene 2—**HOME OF THE SUN.**

Song—"Hold Me In Your Loving Arms"..... Helen Rook and Chorus

Scene 3—**THE SILVER FOREST.**

Song—"My Zebra Lady Fair"......... George White and Zebra Girls

Scene 4—**THE CATSKILL MOUNTAINS.**

Dance.. Dwarf Girls
Ralph Van Winkle, Jr................................ Carl Randell
Song—"I Can't Do Without Girls".... Carl Randell and Country Girls
Rip Van Winkle.. Leon Errol
Jennings B. Ryan.. Will West
Nut Sundae.. Ed. Wynn
Trio—"Twenty Years Ago"..... Leon Errol, Will West and Ed. Wynn

Scene 5—**BARKER'S JUNGLE.**

Dance—"The Birth of a Chicken"................ Anna Pennington
O. Shaw Androcles.. Bert Williams
Prof. Alsoranville Barker.. Will West
The Lion.. Phil Dwyer

Scene 6—**RADIUMLAND.**

Song—"My Radium Girl"........... Chas. Purcell and Radium Girls

Scene 7—**COMMOTION PICTURE.**

Director .... Ed. Wynn
Merry Pickem.. May Murray

Scene 8—**ACROSS THE CONTINENT.**

Song—"Hello 'Frisco" .... Chas. Purcell and Ina Claire

Scene 9—"SOME" MIDNIGHT CABARET.

Al. A. Cart .... **Ed. Wynn**
The Onion Sisters .... **Oakland Sisters**
A Pool Player .... **W. C. Fields**

Scene 10—**AMERICA.**

Major Domo .... **Lucille Cavanaugh**
Dance .... **Cotton Girls**
Navy .... **Anna Pennington and George White**
Army .... **Mae Murray and Carl Randell**

RULERS OF THE WORLD.

Servia—Austria—Russia—Japan—Turkey—Belgium—France—England—Italy—Germany—The United States.
The Wealth of the World—Gold Girl—Silver Girl—Copper Girl—Coal Girl.

Aide to the President .... **Olive Thomas**
Consort .... **Chas. Purcell**
Dove of Peace .... **Kay Laurell**
Columbia .... **Justine Johnstone**

FINALE.

ACT II.

Scene 1—**A CHRISTMAS EVE FANTASY.**

Santa Claus .... Will West
Cinderella .... Mae Murray
Nightie Girls. Olive Thomas, Dottie Wang, Helen Barnes, Nancy Wallace
Sprites .... Marcelle Earle, Margret St. Claire, Dorothy Godfrey, Margaret Morris
Christmas Trees .... Gladys Feldman, May Paul, Edith Whitney, Gladys Loftus
Song—"I'll Be a Santa Claus to You" .... Will West

Scene 2—**HOME OF THE SUN.**

Himself .... W. C. Fields

Scene 3—**HALLWAY OF THE BUNKEM COURT APARTMENT.**

Thomas, the hall boy .... Bert Williams
Sammy, a messenger boy .... Anna Pennington
Adam Fargo .... W. C. Fields
Lotta Pep .... Lucile Cavanaugh
A Tenant .... Lottie Vernon
A Waiter .... John Ryan
Gladiolo .... Justine Johnstone
Constant Bunn .... Leon Errol

Scene 4—**BELASCO THEATRE.**

Song—"Marie Odile" .... Ina Claire

Scene 5—**THE SILVER FOREST.**

Song—"A Girl for Each Month in the Year,"
Chas. Purcell and Month Girls

Olive Thomas, as January; Claire Bertrand, as February; Flo Hart, as March; Mae Paul, as April; Justine Johnstone, as May; Gladys Loftus, as June; Margaret St. Clair, as July; Nany Wallace, as August; Vivian Oakland, as September; Edith Whitney, as October; Gladys Feldman, as November, and Reta Bates, as December.

Scene 6—**A FIFTY-SEVENTH STREET SHOP.**

Flirtation Medley Dance .... Anna Pennington and George White

Scene 7—**BILLIE BURKE.**

Ed. Wynn .... Ed. Wynn

Scene 8—**ELYSIUM.**

"Oriental Love" .... Mae Murray, Lucille Cavanaugh and Carl Randell
The Lady on the Wall .... Kay Laurell
Dance Egyptienne .... Leon Errol and May Hennessey

---

Scene 9—**HOME OF THE SUN.**

Songs { "I'm Neutral" .... / "In the Evening" .... } Bert Williams

Scene 10—**ZIEGFELD DANCE DE FOLLIES.**

The Bcbelin Tapistry .... Kay Laurell
Dance .... Lucille Cavanaugh and Carl Randell
Mrs. Vernon Castle .... Ina Claire
Song—"The Midnight Frolic Glide" .... Helen Rook and Chorus

FINALE.

Musical Director .... Frank Darling

---

The program for the 1915 *Ziegfeld Follies*, in which W. C. Fields appeared with Ed Wynn, Ann Pennington (misspelled "Anna"), Ina Claire, Bert Williams, Leon Errol, George White, Justine Johnstone, and Mae Murray.

W. C. Fields demonstrates an instrument of his own invention, the "kadoola-kadoola," in the 1923 hit, *Poppy.* *

go home. My father and I didn't see eye to eye on a number of things, so one day I just took a hurried departure."

According to biographers Taylor and Alva Johnston (whose 1935 *New Yorker* profile Fields considered "the best and most authentic" biography to that time), the boy then lived as best he could, sleeping in doorways, boxcars, even lavatories, stealing food from fruit and vegetable stands, and making whatever money he could by cheating at cards. An early predisposition to colds brought on by exposure contributed to the rasping quality of his voice. Constant beatings from street gangs resulted in a permanently swollen nose. The hardships Fields experienced from living by his wits—including periods spent in jail—gave him the tough, cynical, untrusting outlook that he had all his life.

During his teens, Fields took various jobs, including racking balls in a pool hall (where he became, not unexpectedly, a pool shark), delivering ice, and selling newspapers. One evening at a vaudeville show, he saw a juggling act and, as he later recalled, "I decided then and there to become the world's greatest juggler."

To achieve this goal, he first began juggling stolen apples; later he used stolen tennis balls. He devoted several years to perfecting his technique, often practicing up to sixteen hours a day. Aside from juggling balls, Fields tried other, at times painful, variations such as balancing a stick on his big toe, tossing it into the air, and catching it again on his big toe. Once he felt sufficiently confident in his ability, he got a job at a fairground in Norristown, Pennsylvania. He left after about a week when he discovered that it cost him more money in fees and transportation than he was paid. His second job, at an Atlantic City pier, combined juggling with drowning. That is, when business was slow at the concession where he worked, his task was to go swimming, feign drowning, and be carried back to the pier. With a crowd of sympathetic onlookers now gathered about him, there would be a ready-made audience for the pier's attractions. After doing his drowning bit for two weeks, Fields was off water for life.

For his juggling act, Fields dressed as a tramp in order to avoid the expense of keeping his clothes clean or the bother of changing costumes. He also did it to offer contrast to a celebrated juggler of the time who always appeared in top hat and tails. Even from the beginning, Fields set about being a comic juggler, getting laughs from making mistakes on purpose and, of course, making his feats even more impressive when he managed to do them successfully.

After working in Atlantic City, he was hired by a traveling burlesque company. Realizing that to get anywhere in his profession he would have to do something beside juggling—even comic juggling—Fields now added a humorous monologue. This didn't last long because he had a slight stammer at the time, and he soon began concentrating on being a pantomimist. A vaudeville scout caught his performance, and he was engaged for the Keith circuit with an act consisting of juggling tennis balls and hats, which he would carefully drop and retrieve in a comical manner. He also balanced sticks on his toes and fingers and capped the act with his spectacular finale of juggling twenty-five empty cigar boxes. To vary the routine, Fields introduced his pool-shooting act in vaudeville at the Orpheum Theatre in Boston. At nineteen, he was earning $125 per week.

As a pantomimist, the comedian had one special advantage over other comics: his humor could be appreciated all over the world. After making his European debut in London in 1901 at the Palace Theatre (his billing read "Wm. C. Fields, the Distinguished Comedian and Greatest Juggler on Earth, Eccentric Tramp"), he toured France, Germany, Italy, and Spain and also performed in other countries. He returned annually to Europe and in 1904 extended his travels to include South Africa.

Though Fields continued to enjoy success in vaudeville, he was anxious, as were most variety entertainers, to win applause on Broadway. An opportunity came when James McIntyre and T. K. Heath, a blackface vaudeville team, put together a musical comedy, *The Ham Tree*, and signed Fields to play a character named Sherlock Baffles. What the part had to do with the show's plot—something about two minstrel comics stranded in a small town—has been hard to discover since all accounts of the production emphasized that, even without his tramp costume, Fields did little to change his act. Possibly because there was no other way to make note of it in the program, his specialty was listed under "Musical Numbers." But "A New Way To Play Tennis" was not a song; it was W. C. Fields juggling tennis balls and balancing a tennis racket. *The Ham Tree*, though its initial Broadway run in 1905 was only 90 performances, proved to be a great favorite on the road, and Fields remained in it for about two years.

The comedian made his first trip to the Orient in 1907 and continued touring, with constant changes in tricks and comic business, through 1914. That year, during an engagement in Melbourne, Australia, Fields received a cablegram which, he felt, would finally give him the chance to be accepted as a major attraction in a major Broadway production. The cable was from producer Charles B. Dillingham offering him a part in his next musical *Watch Your Step*. There would be songs by Irving Berlin (his first complete score) and a cast headed by America's greatest ballroom dancers, Mr. and Mrs. Vernon Castle, plus monologuist Frank Tinney and the song-and-dance team of Elizabeth Brice and Charles King. Fields promptly cancelled the rest of his Pacific tour and boarded an American tramp steamer at Sydney for a thirty-nine-day journey to San Francisco. From there he traveled by train to New York, then up to Syracuse, where he joined the cast of the Broadway-bound musical just two days before its out-of-town opening. *Watch Your Step* had a slight story line (the program credit read "Plot, if any, by Harry B. Smith") which was easily forgotten, with most of the show put together as if it were a revue. In a scene in the Automat Department Store, Brice and King introduced "When I Discovered You," then King dropped a coin in the slot of an automatic machine and turned the handle to reveal the beaming, bowing "Silent Humorist," W. C. Fields, his derby tucked smartly in the crook of his left elbow, his right leg balanced daintily on his toes and bent at the knee. After juggling tennis balls, battered silk hats, and empty cigar boxes, Fields acknowledged the applause, then casually glanced down at the fur collar of his top coat and flinched at the sight of a suspicious-looking object. He nonchalantly brushed the "insect" off the fur, blinking with disbelief as it landed on the floor with a loud thud. The noise was provided by the pit drummer who used a pair of cocoanut shells to simulate the sound of the bug galloping off the stage as Fields observed the proceedings with a look of total incredulity.

Fields's performance was hailed by a local newspaper critic as an "excellent offering . . . he does things that any boy from ten to seventy is sure to enjoy." This, apparently, left out too many people to suit Dillingham; notwithstanding the fact that he had summoned Fields halfway around the world to appear in the show, the producer now informed the comedian that the scene was being changed from the Automat Department Store to the Palais de Fox-Trot, and that there was no longer any room for his act. If nothing else, Fields could claim the unenviable distinction of having traveled the longest distance on record to play a one-night stand.

But something else did come of his aborted appearance in *Watch Your Step.* In the audience at the show's Syracuse opening was Florenz Ziegfeld's assistant, Gene Buck, who prevailed upon Ziegfeld to engage the comedian for the *Ziegfeld Follies of 1915*. Thus, after a vaudeville tour, Fields joined Ed Wynn, Ann Pennington, Mae Murray, Bert Williams, Ina Claire, Leon Errol, and George White as the featured cast members when the revue opened at the New Amsterdam Theatre on June 21, 1915.*

Fields acted in one skit (along with Pennington, Williams, and Errol), but his specialty was the pantomime pool game that he had been performing in vaudeville. After ambling on stage, he blinked at the audience, as if surprised to see anyone watching. First, Fields selected a cue at random. He tried it out on the table but when it did not glide smoothly on the felt top, he held it

*For this and for his appearances in the next two *Follies*, the comedian was alternately billed as W. C. Fields and William C. Fields.

closely to his eyes the better to determine that the ridiculously warped stick was not exactly straight. As a further test, he vainly tried putting the tapered end through a loop formed by his left forefinger and thumb. With a shrug, he chose another cue, this time unwarped, but again found it difficult to hold the elusive tapered end steady enough to be chalked. After showing his ineptitude by tearing a hole in the table top and hitting a ball so hard it sailed off the table, he ended the demonstration by managing a whack of such accuracy that all fifteen balls flew in different directions to land in all six pockets. All the while, Fields was the epitome of the consummate pool-hall athlete, prancing and dancing around the table, twisting his body for the proper playing positions, brandishing his cues, and squinting to study the lie of the balls with his chin almost on the table top.

Fields's appearance in this *Follies* was not without a memorable altercation. During the post-Broadway tour, while the show was playing Boston, Fields became aware that his pool-table routine was getting laughs in all the wrong places. Causing this was Ed Wynn, who somehow had thought it would be great sport to hide under the table making funny faces and gestures and, in general, ruining the act. Fields, though seething, continued as if nothing was happening until Wynn popped his head up; with that, Fields brought his cue down hard on Wynn's head, knocking the comic almost unconscious. Believing that this was part of the show, the audience roared, but Wynn had learned his lesson. Though his comedy frequently included performing in other people's routines, he never again attempted to get laughs when he wasn't supposed to.*

His success in the 1915 *Follies* resulted in Fields being signed for the next year's edition, along with Will Rogers, Ann Pennington, Ina Claire, Marion Davies, Bert Williams, and Fanny Brice. Fields juggled and did a mad burlesque of a croquet game, though this was not quite so well received as the pool-game routine. Fields also did reasonably funny takeoffs of Secretary of the Navy Josephus Daniels and Teddy Roosevelt in a living version of the Hearst newspaper supplement, "Puck's Pictorial."

For the 1917 *Ziegfeld Follies*, Fields had to keep up with such company as Bert Williams, Will Rogers, Eddie Cantor, Fanny Brice, and Walter Catlett. It was in this edition that he introduced his pantomime tennis routine, which revealed a far more physical and farcical side of Fieldsian comedy than was later in evidence. The scene opens showing a backdrop of a wire fence with a tennis net running downstage about four feet from stage left, thus allotting most of the performing area to the right side of the net. Walter Catlett

*W. C. Fields performed his pool-game act in his first film, a one-reeler called *Pool Sharks*, released in 1915. He repeated the routine on Broadway in *Ballyhoo* (1930), and in three feature films: *Six of a Kind* (1934), *The Big Broadcast of 1938*, and *Follow the Boys* (1944).

enters from stage right, eats a banana and drops the peel on the court. Fields enters from stage left, a look of smug confidence on his face and a dozen tennis rackets under his arms. He slips on the banana peel, the rackets go sprawling, and he tries desperately to maintain his composure while attempting to pick up the rackets without losing his hat. Eventually, he has retrieved all twelve and he places them on the edge of a bench up stage near the tennis net.

Catlett, Fields's partner, sits down on the other side of the bench opposite the rackets; his weight tilts the bench and the rackets fly through the air hitting him as he falls to the ground. More business about retrieving rackets. Fields vigorously swings one of them in an effort to impress his two female opponents, Peggy Hopkins (later known as Peggy Hopkins Joyce) and Allyn King, who have entered from stage right and cross over to the mostly hidden side of the court on stage left (actually they simply walk into the wings and are unseen during most of the sketch). The game now begins with Fields serving. After the ball rebounds from the wings (looking, of course, as if it had been returned by either Hopkins or King), a volley continues until the ball knocks Fields's hat off. Expecting to be hit again, Fields wards off the impending blow by raising his left arm in a protective gesture and hunching his neck into his shoulders, but the ball is not returned. He gets a new racket with a handle approximately two feet long and holds it near the strings with the handle sticking straight up behind him. When his hat falls off he tries to put it back but has trouble finding his head and the hat ends up on top of the handle.* Trying to cover his embarrassment, Fields balances the racket on his head. When it falls off, he catches it on the toes of his right foot and balances it. Then, using thumb and forefinger, he nonchalantly raises the racket by its strings, draws his right foot away and balances the racket on the toes of his left foot. He kicks the racket, it turns half a somersault, and he catches and balances it on his right toes again.

After a volley, the ball is caught on the brim of Fields's hat; it drops, bounces, and he hits it with his racket. Another return and the ball is again caught in his hat, but this time Fields lets it roll off and drop into his right hip pocket. Catlett returns the ball after a serve, and he and Fields dash madly from the baseline to the net and back again in anticipation of a slashing return—but the ball merely rolls under the net from the left side. Fields takes a new racket, which has a handle that bends in the center, and as he hits the ball the strings loosen and cover the ball like a drooping fish net. Another racket is chosen and there is another volley until the ball hits—and lodges in—Fields's neck. For a moment he has no idea where it is; then, as he lifts

*This bit, a favorite of Fields, was usually done with a cane. In 1936, he even did it as Wilkins Macawber in the film version of *David Copperfield*.

his head, the ball falls into his shirt pocket. Fields now serves a ball high into the air and he dashes to the net to await its return. But it doesn't come back. After a moment or two of Fields shading his eyes as he squints into an imaginary sun, Miss Hopkins casually walks over to the net and hands him the ball, for which she receives a disdainful smile of thanks.

Now Fields becomes a true virtuoso of the court, slashing the ball with elaborate forehands and backhands. A volley suddenly ends with dozens of balls hurled at Fields and Catlett, knocking them both down. Then, suddenly aware that there are people laughing at them, the men jump up and end the scene by furiously whacking the balls into the audience.

Another sketch featuring W. C. Fields was one of the rare occasions in his stage career that found him involved with liquor. By the summer of 1917, the country was extremely concerned over the Prohibition Amendment to the Constitution that Congress would soon submit to the states for ratification. In the sketch, audiences were given a look at what might occur on a New York street corner once the amendment became law. The specific site shows the maze of scaffolding and crosswalks used for the construction of the Broadway subway line where, on an elevated clearing, stands W. C. Fields behind his whistling peanut roaster. After a policeman has strolled by, a stranger (played by Walter Catlett) furtively approaches the street vendor.

FIELDS: Seeking sustenance from peanuts, friend? Those small yet succulent morsels of tastiness?

CATLETT: Are you Harry?

FIELDS: Harry? I? What prompts you to ask if I am Harry?

CATLETT: I'm a friend of Charlie Bates. He said just to mention his name.

FIELDS *(groping):* Bates? Bates? Charlie Bates? Christened Charles, I presume?

CATLETT: Yes, sir. Of St. Joe.

FIELDS: Of St. Joe, you say?

CATLETT: Yes, sir. He was here last month. Charles G. Bates.

FIELDS: Ah, Charles G. Bates! A bell seems to tinkle. I concede that the name might in truth be familiar. Perchance you too have a monicker?

CATLETT: Yes, sir. Gus Ferderber. *(He offers a calling card, the surface of which Fields thumbs suspiciously.)*

FIELDS: You'll pardon me, I trust. Mountebanks could easily have such things printed in order to fleece honest merchants in the goober trade.

CATLETT: But I *am* Gus Ferderber. I just got to town today.

FIELDS *(doubtfully):* I see. And where is your permanent abode?

CATLETT: St. Joe.

FIELDS *(cynically):* Come, come now. Don't tell me that there are *two* people in St. Joe.

CATLETT: Honest, I've lived in St. Joe all my life.

FIELDS: There's no need to whimper. Do you have anything to support your preposterous statement? *(The stranger removes his straw hat and lets Fields examine its inner band.)*

FIELDS *(reading):* "Joe Zilch, Gents' Furnishings. Paris, London, and St. Joe." *(He returns the hat and offers his hand.)* 'Tis clear you have not told a tissue of lies. I welcome you to our little settlement. Any friend of brother what'shisname—what *was* his name again?

CATLETT: Charlie Bates.

FIELDS: To be sure, to be sure. Old Charlie Bates. As I was about to say any friend of Charlie Bates is a friend of Harry Musgrove Brandywine, Third. Just a second, doc. *(Fields presses a lever and the peanut roaster is transformed into a fully stocked bar, with a brass rail, sawdust on the floor, and an appropriate painting of a nude woman.)* Name it, brother.*

In the 1918 *Follies,* Fields was joined by Eddie Cantor, Marilyn Miller, Ann Pennington, Lillian Lorraine, and Will Rogers. He acted in a sketch about a patent attorney (with the Fieldsian name of Bunkus Munyan), did his juggling routine, and appeared in "A Game of Golf," his latest encounter with a popular sport. One major difference was that Fields had now added dialogue to his pantomime sketches. This exchange reveals the prevaricating nature of a typical Fields character, Col. Bogey:

BOGEY: In the early days in the Canaries we used to tee up on one island and drive to another.

GIRL: How far is it from one island to another?

BOGEY: Oh, about four or five miles.

GIRL: Really?

BOGEY: We used to putt a quarter of a mile. 'Course we had to have the wind behind us.

In the scene, set on a golf course, Fields first complains about a club with a shaft so flexible that it bends around him when he swings ("Caddy, there is too much whip in this club!"). He proceeds to chalk another club like a billiard cue, hit a ball into his caddy's hat, and step on the head of a club which flies up and hits him. Moreover, he is distracted by the sound of his caddy's

*This dialogue was recalled by Marc Connelly in his autobiography, *Voices Offstage.* Fields also included a small portion of the scene in the 1925 film, *Sally of the Sawdust.*

squeaking shoes and by a sticky piece of his caddy's pie that clings first to his shoe, then to his club and finally to his glove. As he is about to address a ball, a beautiful girl saunters by leading a Russian wolfhound. Tipping his cap, Fields comments, "Fine looking camel you have with you." A horsy woman steps on a club and breaks it. A young man comes over with a dead rabbit and throws it into Fields's golf bag. Suppressing his rage, the comedian admonishes him with, "You are a great hulking boy!" A hunter enters, looking up in the air, fires his gun, and a prop turkey falls on Fields's head, provoking the matter-of-fact reaction, "I wish those aviators would fasten themselves in more securely." The scene ends with a dog running on stage, stealing a ball, and everyone in mad pursuit.*

Along with Fanny Brice, Fields was in the *Ziegfeld Midnight Frolic* in 1919, and the following year in both *Ziegfeld Girls of 1920* (a new name for the *Ziegfeld Nine O'Clock Revue*) and also the *Ziegfeld Midnight Frolic*. These were cabaret entertainments performed atop the New Amsterdam Theatre, where the *Ziegfeld Follies* was then playing. Fields juggled in all three shows and performed his croquet routine in both the early and late revues offered in 1920.

The lineup in the 1920 *Ziegfeld Follies* included Fanny Brice, John Steel, Charles Winninger, Moran and Mack, Ray Dooley, and Mary Eaton. For this edition, Fields abandoned his sketches about the sporting life in favor of one dealing with the hazards of driving an automobile. In "The Family Ford," Fields and Fanny Brice played Mr. and Mrs. Fliverton, and Ray Dooley was their bratty daughter, Baby Rose. Along with Mrs. Fliverton's father and a friend they are seen motoring down a country lane with Fliverton at the wheel. Just as the car gets to stage center, it wheezes and stops dead. Fliverton's efforts to restart it, including cranking, only result in the automobile's backfiring. Then the self-starter works but the engine conks out. Next the car starts of its own accord but again it stops dead. When Fliverton orders everyone out, Baby Rose has to be dragged kicking and screaming and she angrily pulls off one of the doors. Fliverton raises the hood to look at the car's engine, but his attention quickly turns to repairing a punctured rear tire. He jacks up the car, replaces the tire, and steam comes out of the radiator. With all the passengers back in the automobile, Fliverton pulls the jack from under it. As the car sinks lower and lower, the riders are thrown onto the stage with all the tires falling on top of them.

In the 1921 *Follies*—whose cast included Fanny Brice, Raymond Hitchcock, Ray Dooley, Van & Schenck, and Florence O'Denishawn—W. C. Fields

*Variations on this routine were performed on the screen in two shorts, *The Golf Specialist* (1930) and *The Dentist* (1932), and in the feature-length *Big Broadcast of 1938*.

made his first Broadway appearance without juggling. He did, however, discover a new irritant: the New York subway. In his sketch, "Off to the Country," which takes place on a subway platform, we again meet the Fliverton family (Fields, Brice, and Dooley) plus the Fliverton son nicknamed Sap (played by Raymond Hitchcock). They have come to board a train that will, presumably, take them to a rural area such as the Rockaways, and in preparation for their outing they have brought along all sorts of impedimenta—a mandolin, a parrot cage, a rifle, fishing poles, a ukulele, an umbrella, tennis rackets, plus various boxes, bundles, and suitcases. Pop Fliverton's first mishap is slipping on worms that have escaped from a can carried by his daughter. Hearing the deafening roar of an oncoming train, the Flivertons prepare to board, only to be a second too late and the doors close in their faces. When, after two more attempts, they finally get inside a car, they are hauled out by a guard claiming that Pop hadn't put the ticket in the box upstairs. By the time Baby Rose finds the tickets and Pop rushes upstairs to put them in the box, another train has gone by. As the family waits disconsolately on the platform, a man hesitantly walks down the stairs, pleads with Pop for brandy, and faints. Pop quickly gets out the brandy and puts it to the stranger's lips—only to have the man flash his badge and arrest Fliverton for violating the Volstead Act.

In addition to performing his own material, Fields also appeared in sketches that he had not written. One, by Channing Pollock, had Fields playing John Barrymore, Fanny Brice playing sister Ethel, and Raymond Hitchcock playing brother Lionel.*

With the 1921 *Follies*, Fields received some of his most favorable reviews to date. As Jack Lait wrote in *Variety*, "Fields addled along through the show, appearing wherever a spot was possible, unctuous, amusing, always the high-grade jester, always helping the book laughs, always feeding as well as partaking. He proved a tower of strength in the last department."

Because of a fight with Ziegfeld, Fields changed allegiance the following year to join *George White's Scandals*, which White had initiated in 1919 as an annual rival to the *Ziegfeld Follies*. Fields's contributions were not among his major efforts. He did a city version of "The Family Ford—here called "Terrific Traffic"—with Winnie Lightner playing his wife. He returned to juggling in another scene, and even revived his take-off on sports in another. This one, about baseball, was considered neither original nor particularly funny.

W. C. Fields had been appearing regularly in Broadway revues—plus the Ziegfeld cabaret shows—for about nine years. Just as he had once been anxious to break out of vaudeville and became part of the so-called legitimate theatre, he was now anxious to break out of revues and into book musicals. Since

*For a more detailed account of this sketch, see page 11.

a leading role in such a production would give him the opportunity to portray a sustained character in a story, he reasoned that it would also give him greater artistic recognition and possibly open a new career. His chance came in 1923 when producer Philip Goodman handed him the role of Professor Eustace McGargle in *Poppy*. Though Dorothy Donnelly was credited with the book and most of the lyrics, she had nothing to do with writing Fields's part. This was the work of Howard Dietz, with an assist from Fields, based on a character in a German operetta that Dietz had adapted for Goodman but which was never produced.*

Madge Kennedy, a popular stage and screen actress in her only Broadway musical, was starred in *Poppy*, and Fields was featured, but the comedian scored such a hit that eight months after the opening, when Miss Kennedy was replaced by her understudy, Fields received solo star billing above the title. (After Miss Kennedy returned to the cast, however, she and Fields were co-starred for the tour.) Fields's part of the rodomontade carnival grifter, juggler, shell-game artist, card shark, and patent-medicine hawker set the pattern for most of his subsequent characterizations. Who else could respond to accusations of forgery, abduction, and other crimes by haughtily saying, "This is no time for idle twitting"? Even his costume—tall white stovepipe hat, fawn-colored coat with huge buttons and a beaver collar and cuffs, flowing ascot, white cotton gloves, checkered trousers, and spats—became something of a sartorial trademark.

The story of the musical, set in 1874 in a Connecticut town, had to do with McGargle's attempt to pass off his foster daughter, Poppy, as a local heiress—only to discover that she really is an heiress. The plot was loose enough to allow Fields to perform some of his specialties and to introduce his poker-game routine. In the scene, he condescendingly agrees to join a group of sour-faced players and, without putting up any money, manages to deal himself a hand that includes four fours, thus winning a $1000 pot.

Fields's ornate speech in *Poppy* caught the proper spirit of carnival fakery and did much to influence the comedian's dialogue in subsequent appearances. Early in the play, Prof. McGargle appears in the carnival midway to deliver a come-on for his shell-game. "This is not a game of chance," he insists, "but a game of science and skill." Suddenly spotting the dignified town mayor coming his way, McGargle quickly removes the shells from his table and replaces them with a Bible. Adopting a preacher-like demeanor, he solemnly intones, "Gambling is the root of all evil. For years I was a helpless

*Dietz later shed his anonymity when, as lyricist, he joined composer Arthur Schwartz to write songs for such revues as *Three's a Crowd*, *The Band Wagon*, and *Inside U.S.A.*

pawn in the toils of Beelzebub." Then turning to the mayor, he inquires, "Perchance, sir, you have read my book on the evils of wagering?" When the mayor shakes his head, the professor appears surprised. "You haven't? It has a blue cover. Perhaps that may recall it to your mind."

Poppy was a great triumph for Fields. The newspaper critics, while admiring Miss Kennedy, offered little doubt that he was the main attraction. "His jaunty and shameless old mountebank has the flavor of someone astray from one of Mark Twain's riverboats or one of Mr. Dickens' groups of strolling players," wrote Alexander Woollcott in the *Herald.* And from Alan Dale in the *American:* "Miss Kennedy was the star of the evening, a beautiful star, but the better-than-star of the evening was W. C. Fields. Mr. Fields was so tremendously funny and so outrageously clever, that he carried everything away from everybody. It was a W. C. Fields evening with a vengeance."*

Having proved himself a character comedian fully capable of carrying an entire musical, Fields was somehow unable to take advantage of his newly recognized ability. His chance should have come with his next show for Ziegfeld. Writer J. P. McEvoy had devised a highly original topical revue, *The Comic Supplement,* with scenes offering a satirical, cartoon-strip view of life in the United States. Fields and Ray Dooley were signed for the leads, but despite the sponsorship and the stars, the show had such an unpromising tryout early in 1925 that Ziegfeld closed it before the scheduled opening. Unwilling to lose everything, however, the producer added Fields and Dooley to the then running 19th edition of the *Follies,* replaced about half the sketches and songs, and added the line, "LATEST 1925 EDITION" to the title. Among the new sketches were four—'A Back Porch," "Drug Store," "Joy Ride," and "Picnic"—that had been in *The Comic Supplement.*

"A Back Porch" (which, with changes, showed up in the 1934 Fields movie, *It's a Gift*) is concerned solely with Fields's attempt to sleep in a hammock on his porch in the face of countless interruptions. Among them: a newsboy with squeaking shoes ascending and descending the stairs leading to the floor above; the collapse of the hammock; a girl yelling to neighbors that there is a telephone call for them in the drug store; an alarm clock ringing on a nearby table; Fields's baby crying so loudly that papa has to stuff bed covers in its mouth; the baby hitting Fields with a mallet; a fruit peddler calling, "Ripe tomatoes, onions, potatoes"; a scissors grinder announcing his trade and ringing his bell; a gun exploding; a policeman's motorcycle engine backfiring; an iceman shouting and banging loudly on the door with his tongs. By this time Fields's pa-

*Fields appeared in two screen versions of *Poppy:* D. W. Griffith's silent version in 1925 (called *Sally of the Sawdust*) and Paramount's 1936 sound film.

tience is at an end and he goes down the stairs to pick up a block of ice and put it in the ice box. But the lid jams on his fingers, and in anger he drops the ice into the nearest trash can.

The other durable sketch, "Drug Store," also found Fields as the victim of numerous irritations as customers troop in, not to buy anything, but to use his telephone, have specks taken out of their eyes, request the correct time, use the ladies room, and purchase a stamp in order to get a free gift. (This scene became the basis for Fields's 1933 film short, *The Pharmacist.*)

When Fields could find no suitable theatre vehicle after the *Follies*, he began a new career in silent films. None of his eight movies was successful. In April 1928 he complained to his wife in a letter, "I am at a stage where I cannot get an offer." Fields's lack of prospects at the time turned out to be a boon for showman Earl Carroll. Having recently been released from prison, where he had served time on a morals charge, Carroll was anxious to redeem his reputation with a smash hit. The means for this redemption would be another edition of his *Earl Carroll Vanities*, a glitzy, frequently tasteless annual revue that he had offered between 1923 and 1927. The main difference was that this time, Carroll realized, he desperately needed a prestigious star attraction. An attempt to secure the services of Beatrice Lillie proved fruitless, but after protracted negotiations that included stiff financial terms and a demand for solo star billing above the title, W. C. Fields agreed to appear in the show (thus making him the only major performer ever to have been in revues offered by the three leading impresarios in the field, Ziegfeld, White, and Carroll).

The *Vanities* did turn out to be a smash and, since Fields had more to do in it than he had had in any other revue, a large credit for the success belonged to him. At least three of his sketches were singled out for praise. In "Stolen Bonds," he played an Alaska gold prospector—bundled up in mackinaw, tippet, and fur hat—who muttered " 'Tain't a fit night out for man or beast" every time he opened his cabin door and was pelted by a fistful of fake snow. "The Caledonian Express" was an especially inventive bit in which Fields, using a Scottish accent, appeared in four roles—a train guard, a conductor, a station master, and a policeman—as each was summoned to evict a gentleman who was smoking in a non-smoking compartment. After Fields's first appearance, each succeeding appearance was accompanied by an actor wearing a mask of the railway employee that Fields had just played. Thus, at the end when he came in as a policeman, Fields was also accompanied by lookalikes dressed as a guard, a conductor, and a station master.

In the Fields sketch, "At the Dentist's," the comedian appeared as Dr. Pain, whose main interest is golf and who is equally disdainful of his patients and members of the medical profession ("There's a doctor in this building who's been treating a man for yellow jaundice for fifteen years and only yesterday found out he was a Chinaman"). In one sequence he tussles with a patient (Ray Dooley) to force her to sit still for an extraction. As he grapples with her, he somehow tears a window curtain, which makes him grope nervously thinking he's ripped his pants. Another patient, Mr. Foliage, sports whiskers so thick that Dr. Pain must cut them with a pair of scissors in order to find his mouth. After further examination, a bird flies out of the beard and the dentist puts on his hunting cap and pokes a shotgun through the thicket. Backing away, he tosses a golf ball into the growth as if to scare a flock of birds. This is too much for Foliage, who dashes out of the room vowing never to return. Dr. Pain's parting remark: "I'll set a mousetrap for you!"

When the *Vanities* went on tour, Fields added his bootlegging sketch from the *Ziegfeld Follies of 1917* and his golf sketch from the *Follies of 1918*.

After scoring a success in the *Vanities*, Fields returned to Broadway late in 1930 in *Ballyhoo*, a book musical based on the exploits of one C.C. Pyle, who, a few years back, had sponsored a cross-country foot race called a "bunion derby." In the show Fields played the part of Q.Q. Quayle—but by any name he was easily recognized as a modern incarnation of that supreme con artist, Prof. Eustace McGargle. Despite Fields's efforts (including resurrecting his pool-table routine, his juggling routine, his breakable motor-car routine, his drugstore routine, and his poker-game routine), the show itself was such an incoherent mess that it ran only a little more than two months. And that was only because producer Arthur Hammerstein bowed out, and the principals took over the show as a cooperative venture with a 50 percent cut in salary. Though the critics failed to be taken in by the entertainment, Fields remained a great favorite. According to Brooks Atkinson in the *Times*, "Life is a serious business for this mountebank. Pondering on the fickleness of fate, he champs his unlighted cigar, swings his cane thoughtfully and bulges at the waistline. A masterly dignity surrounds his skittish grey derby. The eyes burn with the fierceness of the practical philosopher. Although his fiancial affairs are in a desperate state, he is a man of principle. He will not rob a sleeping cowboy who carries a loaded pistol. 'A thing like that would be dishonest,' he says in a scornful tone as he scrambles clumsily through the furniture to get out of the way."

February 21, 1931, marked the final Broadway appearance of W.C. Fields. With his prospects at a low ebb, he returned to the chancy world of moviemaking. After a shaky start, by the middle of the 1930s he so dominated the medium that he achieved the near-legendary status of a Chaplin or a Groucho Marx. With Groucho, too, Fields shared a cynicism and perversity of character that were not exactly the qualities one would have expected to win success in a medium geared to wholesome family entertainment. Yet with such films as *Poppy, David Copperfield, Never Give a Sucker an Even Break, The Bank Dick,* and *My Little Chickadee,* he managed to establish his unique personality on the screen as few comics before or since. He died in Pasadena, probably in his sixty-sixth year, on Christmas Day, 1946.

Thanks to film, Fields performances have been preserved so that they may continue to be enjoyed throughout the world. But it cannot be overlooked that it was on the Broadway stage that he first developed the special character that has made his screen contributions so outstanding—a bitter, frustrated, put-upon, unbowed rascal whose pride never wenteth before or after a fall.

Willie Howard

W. C. FIELDS COULD BLUSTER his way out of adversity; Jimmy Durante could explode against it; Bert Lahr could face it by bellowing and grimacing; and Victor Moore could wear it down with childlike innocence. But for the constantly put-upon Willie Howard, the usual recourse was little more than a philosophical shrug coupled with a look of total perplexity. Indeed, most of the time he looked like someone who found it hard to cope with anything, good or bad, and his stoop-shouldered stance and worried expression gave him the appearance of a perpetually confused little man buffeted by countless injustices.

There were times, though, when Willie was allowed to hit back with the weapon of mischief, scampering his way through escapades that found him eventually gaining the upper hand in contests with overbearing figures of authority. In this, of course, he was aided by his harassed appearance that cloaked his true intent. Though Willie looked like a nebbish, he never sought pity or sympathy. Yet all his leering, ogling, and wisecracking could not hide that plaintive quality which was so much a part of his personality. We knew he could take care of himself but we were never sure that he knew.

Willie's face was more suitable for a mask of tragedy than for a mask of comedy. All the lines turned down. His thick, black, often unruly mop of hair crowned and fell over a pale, craggy face, with expressive sad eyes and a wide, drooping mouth. The focal point was unquestionably his nose, a promontory that descended from a hook high on the bridge straight down to a point. His only physically comic feature was his frequently wiggling jug ears.

Despite his woebegone appearance and early indications that he harbored an ambition to become a dramatic actor, Willie remained a Broadway clown throughout his life. There were brief forays into radio, movies, recordings, night clubs, even television, but these neither lasted long nor did they produce any memorable comic triumphs. From 1912, his first year on Broadway, virtually to his death in 1949, scarcely a season went by without a stage appearance by Willie Howard, either in New York or on the road. Only a terminal illness prevented him, at sixty-two, from starring in what would have been his twenty-fourth Broadway show. After opening in *Along Fifth Avenue* in New Haven, he had to relinquish his part to Jackie Gleason when a liver ailment put him in the hospital. He died on January 12, 1949, just one day before the show's New York premiere.

Throughout most of his career, Willie adopted a Yiddish accent. Unlike Fanny Brice's, which was largely used as a comic contrast to the parts she played, Willie's accent was often used to emphasize his ethnic identification. The comedian was, however, a master of other dialects as well—mainly French, Spanish, and Scottish—which he would combine with his Yiddish trademark. Two other talents: he was a superb mimic, seldom appearing on stage without impersonating leading actors and entertainers of his day, and he was an accomplished singer. It was because of his singing ability that he developed his celebrated operatic take-offs, including the *Rigoletto* Quartet.

Until 1940, all his appearances in revues found Willie teamed with his older brother Eugene, who served as his straight man. Eugene was balding, deadpan, and pompous, and together they offered the classic contrast of the seemingly defenseless little fellow getting the verbal best of his domeering, self-assured partner. At first the team was known as Howard and Howard; later they were usually billed as Willie and Eugene Howard. The act broke up in 1940 when Eugene retired to become his brother's business manager and gag writer.

Willie Howard, né Levkowitz, was born in Neustadt, Silesia, on April 13, 1886, the son of a cantor. The family emigrated to New York from Germany when he was a child and settled in Harlem. As a boy, Willie was an undisciplined prankster, constantly getting into trouble at school, and it wasn't long before he was expelled for disrupting classes. With his natural comic gift, he had no other thought than to become an entertainer, and he was soon picking up nickels and dimes clowning and singing in amateur shows around the city. His first more-or-less regular job, at fourteen, was as a song plugger for publisher-songwriter Harry Von Tilzer. Willie's duties were to attend vaudeville performances at Proctor's 125th Street Theatre, where singers would perform the latest Von Tilzer creations. Following each rendition, Willie would rise from his balcony seat as if spontaneously inspired and, after apparently only one hearing, lustily deliver a second chorus of the song. His vocal prowess soon caught the attention of a rising young producer named Florenz Ziegfeld, who gave the teenager his first legitimate theatrical employment. But not for long. Willie was hired to sing from a theatre box during the production of *The Little Duchess*, starring Ziegfeld's wife, Anna Held. When the youth

Willie and Eugene Howard as they appeared in a sketch satirizing the musical, *Mecca*, in *The Passing Show of 1921.**

Ginger Rogers and Willie Howard in a scene from the Gershwin brothers' 1930 musical, *Girl Crazy.**

The sheet music cover (designed by Russell Patterson) of *Ballyhoo of 1932*, which featured Willie Howard, Jeanne Aubert, and Bob Hope.

Willie Howard as Prof. Pierre Ginsberg in *George White's Scandals of 1935*.*

Shown on The Playbill cover of the 1939 *George White's Scandals:* Eugene and Willie Howard, Ella Logan, Ann Miller, The Three Stooges, Ben Blue, and Collette Lyons. (Playbill ® is a registered trademark of Playbill Incorporated, N.Y.C. Used by permission.)

began singing on opening night of the show's tryout in Washington, the unsteady sound that emerged revealed that he was undergoing a change of voice, and Ziegfeld had to dismiss him. What really upset Willie was that he had paid for his boarding-house room in advance; rather than forfeit the week's rental money he remained in Washington for the prescribed time, even though there was nothing for him to do.

Once his voice settled down, Willie discovered his talent for vocal mimicry. Teamed with Eugene, he developed a vaudeville act featuring joke-telling, imitations, and close-harmony singing. In their comedy routines, Willie was usually the harried but sharp-tongued bellhop or waiter and Eugene the self-assured but gullible salesman or manager. They succeeded so well in vaudeville that by 1912 they were earning $450 per week.

The Howards' Broadway opportunity came the same year when Lee and J. J. Shubert decided to resurrect the title of a late nineteenth-century revue, *The Passing Show,* and present their own series of revues as a rival to the flourishing *Ziegfeld Follies.* The Shubert offerings, which were all shown at the Winter Garden ("A Music Hall Devoted to the Continental Idea of Varieté"), were less elaborate or artistic than the Ziegfeld shows, but they turned out to be well-attended collections of songs and sketches, mostly satirizing current events. As special attraction, they were decorated by rows of overexposed girls (called, according to height, "show girls," "mediums," and "ponies"), who pranced on a runway that jutted down the center aisle into the audience.

The first Shubert *Passing Show,* which opened on July 22, 1912, carried this information in its program: "A kaleidoscopic almanac, in seven scenes, presenting the comic aspects of many important events, political, theatrical and otherwise, and embracing the sunny side of 'Bought and Paid For,' 'Bunty Pulls the Strings,' 'A Butterfly on the Wheel,' 'Kismet,' 'The Typhoon,' 'The Quaker Girl,' 'The Pirates of Penzance,' 'Oliver Twist,' etc." Besides Howard and Howard, the cast included such comic personalities as Charlotte Greenwood and Trixie Friganza. At the time, Willie and Eugene had not yet gotten past routines that gave them little more than exchanges consisting of easy-to-see-coming setups and comic responses based on misunderstandings of the questions. In the opening scene, a New York pier on which passengers have been debarking from an ocean liner, Eugene, as the customs inspector, interrogates newly arrived Willie:

EUGENE: Where did you come from?
WILLIE: I just got off that boat there.
EUGENE: Well, do you have anything to declare?
WILLIE: I'm all right, I declare.
EUGENE: Where are your papers?

WILLIE: I haven't got the papers but I've got the makings.

EUGENE: Say, you're pretty fresh. Come on, get out your trunks.

WILLIE: Are we going swimming?

EUGENE: Not your swimming trunks. You've got to pay duty on your trunks or you can't stay in the country.

WILLIE: I'd rather stay in the city.

EUGENE: Have you got any relatives here?

WILLIE: Yes, I've got a twin brother two years older than I am.

EUGENE: Is he naturalized?

WILLIE: Well, one of them is a glass eye.

And when Willie revealed that his former occupation had been that of a jockey, that was all the cue needed for him to close the scene to the jazzy beat of Irving Berlin's "Ragtime Jockey Man."

Later in the show, Willie did an imitation of actor David Warfield in his successful play *Peter Grimm* and, in a restaurant scene, singer Jack Norworth. The latter was introduced by this exchange between Willie, applying for a job as a singing waiter, and Eugene as the manager of the establishment who asks about his background:

WILLIE: I was the leader of a circus orchestra.

EUGENE: How many pieces were in the orchestra?

WILLIE: Six pieces—fife and drum.

In the second act, Willie and Eugene joined Trixie Friganza and Ernest Hare in a number called "Metropolitan Squawktette" that was the forerunner of Willie's celebrated *Rigoletto* Quartet routine.

From 1912 through 1923, the Shubert brothers provided the Howard brothers with steady employment in seven revues. Early in 1914, *The Whirl of the World* came to town promoting itself as "The Biggest, Most Splendid, Most Spectacular and Costliest Production Ever Staged at the Winter Garden. An Isle of Gorgeousness, Fun and Music. Entirely Surrounded by Girls. Wherever You Look—Just Girls. Whenever You Look—Just Girls." The show further claimed that it had "The Most Remarkable of All Musical Comedy Casts," including—in addition to Willie and Eugene—Lillian Lorraine, Bernard Granville, Rozsika Dolly, and ballerina Lydia Kyasht.

The Whirl of the World had an easily disposable framework (something about an American who bets the Marquis Tullyrand that he can woo and win thirty girls in thirty days), but otherwise it was little different from the usual revue. Willie again trotted out his imitations of Warfield and Norworth, did another grand-opera burlesque, and told jokes ("What is worse than having an earache and a toothache?" "Rheumatism and St. Vitus Dance").

Willie had a number of opportunities for comic characterizations in *The Passing Show of 1915*. He appeared in sketches as the meek Androcles (an inquiry for Androcles the Christian drew laughs when Willie replied, "Are you looking for me?"), as Trilby to Eugene's Svengali, and as Hamlet to Eugene's Othello. A number called "Broadway Sam" prompted an impression of Al Jolson, and a Hollywood spoof found Willie doing a take-off of Charlie Chaplin in a scene with Marilyn Miller's Mary Pickford. The cast also included singer-comedian Ernest Hare and future opera tenor John Charles Thomas.

The Howard boys—plus Marilyn Miller, Ernest Hare, blackface comics McIntyre and Heath, and monologuist Walter C. Kelly—returned to the Winter Garden in 1916 in *The Show of Wonders*. Though initially there had been an an attempt to give the revue some kind of outdoors theme (the original title had been *Back to Nature*), what emerged was barely distinguishable from the *Passing Show* formula. In one scene Willie appeared as Mendelssohn, and Eugene as Liszt, to protest the way their compositions were being ruined by modern ragtime arrangements. In another, Willie imitated Warfield and Norworth (though the switch here was that Willie did Norworth doing Al Jolson). Early in the show, a sketch called "The Deer Trail" unearthed these feeder-and-gagster exchanges:

EUGENE: A laughing hyena is an animal that eats every three months, and drinks only every six months. And although his wife is in the next cage, he's allowed to see her only every thirteen months.

WILLIE: I'd like to know what he's got to laugh at.

Later, Willie told of his encounter with a lion:

WILLIE: I slammed the door of the cage shut and the lion and I were shut in.

EUGENE: Then what did he do?

WILLIE: What could he do? He had to kill me.*

A strong cast was assembled for the 1918 *Passing Show*, headed by Fred and Adele Astaire (their second Broadway appearance), Frank Fay, Nita Naldi, Charles Ruggles, and the Howard brothers. The boys were in still another operatic take-off, with Violet Englefield as Galli Curci, in which the diva was urged to put a little ragtime into her arias, and Willie expanded his mimicry

*See page 16 for a comparison with Fanny Brice's Baby Snooks line in the 1934 *Ziegfeld Follies*.

repertory to include the Scottish entertainer Harry Lauder. A travesty on Oscar Wilde's *Salome,* the first-act finale, took its lead from the previous Androcles burlesque by deriving humor from the sight of the Semitic-looking Willie appearing as a Christian. In the ornately decorated scene, Salome clamors for the head of John the Baptist which, when brought in on a salver, turns out to be that of the German Kaiser wearing his spiked helmet.

The Passing Show of 1921 opened late in December of the previous year. For the first time, Willie and Eugene Howard were awarded feature billing, with their names in the program twice the size allotted to Marie Dressler and Harry Watson. The Howards' opening scene was a desert locale with a Sphinx painted on the backdrop bearing a head made to look like an uncharacteristically silent William Jennings Bryan. Willie and Eugene made their entrance riding on the back of a two-man camel, which was equipped with a street-car bell and automobile headlights. Eugene, up front, guided their way by tooting on a klaxon, and Willie, seated behind the rear hump, played solitaire. The parched atmosphere represented New York during Prohibition, and the dialogue dealt with various ways to circumvent the Volstead Act (e.g., Willie had a pet snake that bites strangers on the arm, thus requiring them to take a strong antitoxin dose of whiskey which had been hidden in one of the camel's humps).

Other scenes showed more of Willie's range, combining revised versions of well-tested material with some fresh routines that resulted in his most successful *Passing Show* appearance. He did an imitation of Frank Bacon in a scene from his long-running play, *Lightnin'*, that was hailed as almost a facsimile of the original. He adopted an Oriental sing-song as a Chinese character in a scene from *Mecca.* To no one's surprise he did an opera spoof, though for the first time his target was the Quartet from *Rigoletto,* in which he was again abetted by Violet Englefield and brother Eugene, with Ina Hayward making it a foursome. As expected, Willie impersonated Jolson, Norworth, and Lauder, though the Lauder burr was now filtered through a Yiddish accent. Reviewers heaped praise on Willie as they had never done before. For example, Kenneth Macgowan in the *Globe:* "The real feature of *The Passing Show* is Willie Howard. No one at the Winter Garden except, perhaps, Al Jolson, has ever come so near genius as this Howard. His version of Frank Bacon is uncanny in one so thoroughly and racially mannered. But when he sets Harry Lauder singing 'The Saftest* o' the Family' as he would do it on the East Side, Howard is simply colossal. There is art here."

The Howard brothers were joined by deadpan monologuist Fred Allen, song belter Ethel Shutta, and British leading man Arthur Margetson—all making

*Scottish dialect for "Softest"—as in the "softest headed."

their Broadway debuts—in *The Passing Show of 1922*. Again the sibling team was the standout attraction. In a skit set in a phonograph record shop, Willie answers an ad for a salesman and, after a gag-strewn exchange with proprietor Eugene, he is hired. Inevitably, he drops and breaks a stack of records, much to Eugene's consternation. But Willie knows just what to do. When a customer comes in to listen to recordings by Harry Lauder, Al Jolson, Gallagher and Shean, and Eddie Cantor, he hides inside a phonograph console and mimics the singers' voices as Eugene appears to be putting the records on the turntable. In another scene he went back to imitating David Warfield. In still another he refused to attend a christening because he couldn't stand to see a baby get hit over the head with a bottle.

There were two more *Passing Shows* on Broadway, but the 1922 edition was the last in which Willie and Eugene appeared.* The fact that it was the least successful (running only 85 performances) may have had something to do with it, but it was primarily because the brothers felt that the time had come—at least for Willie—to move on to something more challenging. It took almost two and a half years, but early in 1925 the Shuberts at last found a vehicle they felt was suitable. The trouble was that the script offered nothing suitable for Eugene who, for the first time since they teamed up, did not appear with his brother.

Sky High, the musical in which Willie made his solo starring debut, was a somewhat curious affair since it was based on a London musical, *Whirled into Happiness*, which, in turn, had been adapted from a German operetta. Though considerably revised for New York, the show retained its London setting and characters. In the story, Willie played a music-hall coatroom attendant who poses as a lord to win the approval of his girl's parvenu family. In general, the critics were impressed with Willie's ability to carry an entire show. As Gilbert Gabriel wrote in the *Telegram:* "Last night's audience replied to the vigor of his labors with all the usual proofs. He seems intent, when in London, on speaking as the Londoners do, so that the tang of his old accent is half missing. His gnomic, worried, sputterful ways put across even more points than the librettists had tried to sharpen for him; and then he sang, as of old, with a gentle and semi-religious zeal."

The easily ignored book allowed for gags ("What did the fortune teller say after she read your mind?" "She said she enjoyed the vacation"), and toward the end of the second act Willie trotted out some of his most requested imitations (Jolson, Cantor, Lauder). There was even room in the script for an opera take-off. But the comedian seems to have received his biggest laughs

*Late in 1945, the Shuberts tried to revive *The Passing Show* with Willie Howard starred, but the revue closed in Chicago before reaching New York.

when, given the chance to take over for a missing barber, he ties a customer to a chair and inadvertently pokes a lathered brush down his throat. As a result of Willie's drawing power, *Sky High* achieved a Broadway run of six months.

The musical may have been a departure from Willie's accustomed string of revues in that it gave him a chance to play a sustained character, but *Sky High* was hardly a significant breakthrough in his career. Since the Shubert brothers planned no other book musical for Willie—nor did anyone else—he and Eugene felt that they might discover greener fields in a revue form somewhat different from the *Passing Show.* So it was that the reunited Howard brothers next chose to be seen on Broadway under the banner of George White in the eighth edition of his annual *Scandals,* a more youthful, faster-paced, and higher-stepping revue than either the Ziegfeld or the Shubert offerings. The 1926 edition, which opened in June, had a stellar line-up of talent headed by Harry Richman, Ann Pennington, Tom Patricola, and Frances Williams, and a winning score by DeSylva, Brown, and Henderson (including "The Birth of the Blues," "Lucky Day," and "The Black Bottom"). Generally anointed the premiere edition of the series, this *Scandals* made history by charging $55.00 per seat for the first nine rows on opening night. (The top price reverted to $5.50 thereafter.)

Willie and Eugene made the most of their opportunities. As a sentimental gesture, they first appeared in the show reminiscing in song about their years at the Winter Garden. In a sketch called "Phoney Talk" (by Billy K. Wells), Willie, seen at home with his wife at stage right, receives a telephone call from his sweetie (Frances Williams) in her apartment at stage left. To cover his confusion caused by his wife sitting in the same room, Willie makes a date over the telephone by pretending that he is talking to a customer:

GIRL: Are you coming over?

WILLIE: Oh, yes, Mr. Brewster, I'll take care of your order.

GIRL: I need some Scotch.

WILLIE: Scotch plaids? Yes, Mr. Brewster, we have some very fine Scotch plaids in stock.

To the girl's consternation, her husband arrives home unexpectedly, and she now warns her errant lover to keep away by making it appear that she is talking to her dressmaker. Willie's wife leaves the room while he is still on the phone. Upon her return, he finishes the conversation by saying pointedly, "Yes, Mr. Brewster, very well. All right, Mr. Brewster, goodbye." Willie's self-satisfied expression changes abruptly with his wife's announcement, "Mr. Brewster is here to see you." Blackout.

Other scenes offered Willie a wide variety of comic opportunities. In "The

Feud" he appeared as the head of a warring Kentucky mountain family. In "Drama of Tomorrow" he was seen as a titled Englishman whose house is burgled. In "Lady Barber," he was the frightened customer being nicked by barber Frances Williams ("When I want my face lifted," Willie tells her, "I'll take a ride in an elevator"). In "The Good Old Days," he played an old man trying to cope with mechanical conveniences fifty years hence in 1976. From the terminal of the Trans-Atlantic Rocket Station on Times Square, passengers are able to travel by chute all over the world in a matter of minutes, and eat food—in the form of pellets—that is dispersed from slot machines. When a young married couple enters, the husband puts a coin in a machine and a baby comes flying out. This is too much for Willie and, yearning desperately for the good old days, he falls in a faint.

Willie and Eugene were part of the elaborate first-act finale, a confrontation between classical music and jazz that apparently owed its origin to a similar musical clash in the 1916 revue, *The Show of Wonders.* First Willie and Eugene, as Beethoven and Liszt, decry the vulgar music known as the blues, which is then stoutly defended by Harry Richman, explaining its origin in the stirring "Birth of the Blues." The Fairbanks Twins uphold the classics with Schumann's "Traumerei" and Schubert's "Serenade," while the McCarthy Sisters give testimony with "Memphis Blues" and "St. Louis Blues." All opposition crumbles with the playing—and choral singing—of Gershwin's "Rhapsody in Blue." The Howard brothers are now happy to swing open the heavenly gates, while trilling seraphim welcome the once banished blues into Paradise.

In the *George White's Scandals of 1928*, Willie and Eugene were again featured with Harry Richman, Ann Pennington, Tom Patricola, and Frances Williams, and the songs were again by DeSylva, Brown and Henderson. In Billy K. Wells's sketch, "The Ambulance Chaser," the brothers established the same kind of relationship they would later develop in their legendary "Pay the Two Dollars" skit. The scene is a hospital room, and Willie is an accident victim lying in bed with his head, arms, and legs bandaged. Lawyer Eugene barges in, introduces himself, and offers to take the case. As he asks the patient for details of the accident, he keeps interrupting with his own version:

PATIENT: I saw the automobile coming . . .

LAWYER: You didn't see anything.

PATIENT: The chauffeur blew his horn . . .

LAWYER: He did not! Without a word of warning he knocked you down.

PATIENT: When I was knocked down, I jumped up . . .

LAWYER: You couldn't move!

PATIENT: I hollered for help . . .

LAWYER: You were unconscious! Now regarding your wife, when she heard of the accident . . .

PATIENT: She took charge of the store.

LAWYER: She went to bed with a nervous breakdown!

PATIENT: She didn't.

LAWYER: I'll arrange it.

Because there were no witnesses, the lawyer further promises to produce three from Chicago. At the end of the skit, the alleged victim turns out to be even more of a fabricator than the lawyer.

Impresario George White, obviously at a loss for new ideas for the first-act finale, simply offered another classics-versus-popular music songfest. Willie, again upholding "serious" music, bids a tenor render Schubert's "Ave Maria" and Wagner's "O Evening Star." Harry Richman, again the champion of pop music, sings the praises of "A Real American Tune." He then delivers a tribute to, of all people, Victor Herbert, and introduces such curious examples of purely native music as "Gypsy Love Song" (sung by Eugene), "Kiss Me Again," and "March of the Toys."

One scene, though, was especially imaginative. Called "Vocafilm," it first showed a filmed sequence with Harry Richman on a sound screen singing "I'm on the Crest of a Wave." Suddenly, Willie Howard, seated in a theatre box at the right, begins making audible cracks about the singer's talent. Richman abruptly stops singing, and, while still on screen, engages in a heated exchange with Willie. When Richman offers to switch places with him if he thinks he can do any better, Willie leaves the box and is immediately seen singing on the screen while Richman shows up in the box. Now it's Richman's turn to do the heckling. The scene ends with both men on screen together in a duet.

Willie and Eugene were back in the tenth edition of the *Scandals* the following year. This time, with Frances Williams the only other headliner, the comedy burden rested even more heavily on Willie's sagging shoulders. A successor to the 1926 "Phoney Talk"—called "Phoneyfibs"—presented Willie in his apartment with his tootsie while his wife is away on a business trip to Philadelphia. To allay his girl friend's fears that they will be discovered together, Willie telephones his wife, who answers the call seated on her boyfriend's lap ("The meeting was a big success and I'm sitting pretty"). After wifie asks Willie about the health of their baby daughter, she tells him, "You'd better undress her and put her to bed. And I think you'd better sleep with her tonight." To which Willie answers, glancing at his sweetie with a twinkle and a leer, "That's a damn good idea!" Blackout.

In the sketch called "Stocks" (by Billy K. Wells), in which Willie and Eugene played detectives, the alleged purpose was to keep the audience informed about closing stock prices without interrupting the show. The scene is a stock broker's home:

BUTLER: Miss Conda, sir.

BROKER: Anna Conda 110¾?

BUTLER: Yes, sir. She arrived in a Checker Cab 71⅜. *(Girl enters.)*

BROKER: Anna, I told you to stay in Richfield 41⅞.

GIRL: I couldn't Standard Oil 38⅝. I had to come. Montgomery Ward 129¾ found the American Bank Note 148⅝ we forged in his National Cash Register 126½! *(The broker tells the girl his plans for their getaway and she leaves. Two detectives enter.)*

BROKER: Who are you?

DETECTIVES: Abraham and Straus 129⅞. Detectives for the Irving Trust 76¼, Chase National 208½, and Chelsea 99¾.

FIRST DETECTIVE: When people in the United States Steel 202⅜, we always get 'em.

SECOND DETECTIVE: This means twenty years for you in the American Can 164⅞.

In "The Man of the Hour" (by Irving Caesar, Lew Brown, and B. G. DeSylva), Willie got the chance to combine a Spanish accent with Yiddish. In the somewhat macabre episode, only a slight exaggeration of what still occurs in parts of the world today, Willie appeared as Rodriguez Alvarez, the newly elected—and 450th—El Presidente of Mexico. Convinced that the people love him, he makes his inaugural address (unseen to the audience) from the balcony of the presidential palace. As soon as he starts, fruit and vegetables are hurled at him, and gun shots are heard. Even though he's been hit, Alvarez keeps talking about how much the people love him. After a volley of shots, El Presidente staggers back into the room, bandaged from head to foot, and drops dead.

In addition to appearing in sketches, Willie tried out a French-Jewish accent for the first time in a song called "We Americans," in which members of various nationalities sing of their affection for "this grand and glorious new land," despite petty irritations such as one-way traffic, the high price of theatre tickets (they were then $5.50), traffic jams, etc. As for imitations, Willie now added George Jessel to his dependable Cantor and Jolson routines. tines.

Willie's second and unquestionably best-known appearance in a book musical occurred in 1930—though he was not the first actor signed for the comic lead. *Girl Crazy,* the new Alex Aarons–Vinton Freedley production with a score by George and Ira Gershwin, had a book by Guy Bolton and Jack

McGowan with the part of Gieber Goldfarb initially written for Bert Lahr. But George White also wanted Lahr for *his* show, *Flying High;* after Lahr broke his contract with Aarons and Freedley, Willie was awarded the role. It was his second time out minus Eugene, but he did appear in such illustrious company as Ethel Merman (her first Broadway musical) and Ginger Rogers (her second). Merman's songs were "I Got Rhythm," "Sam and Delilah," and "Boy! What Love Has Done to Me!," and Rogers sang "But Not for Me" and "Embraceable You."

Problems occur early in *Girl Crazy* for Gieber Goldfarb, a New York hackie who has driven a fare to Custerville, Arizona. Urged to run for sheriff against the town bully, he wants no part of it when told that the last sheriff had been shot ("I don't look my best by candlelight"). He finally agrees when his supporters assure him that he won't be in any danger as long as he lets the people do as they please. This gives Gieber the idea of treating Custerville the way the police handle things in New York, and he is cheered lustily in song:

TOWNSPEOPLE: So vote for Gieber Goldfarb, he's all right!
So vote for Gieber Goldfarb, man of might!
They needed a man who knows the game
Through serving a lot of time!
Goldfarb!

GIEBER: That's I'm!*

Girl Crazy gave Gieber the chance to wear comic disguises to help him keep out of the clutches of the nasty character he defeated for sheriff. In one scene he appears as an Indian, wearing a loin cloth and war paint, and burdened down by an excessively large headdress. With shoulders hunched and a hammer, rather than a tomahawk, in his right hand, he performs a mournful, skipping dance to the beat of a ragtime Indian melody. When a real Indian speaks to him in his native tongue, Gieber answers in Yiddish, a language that the Indian is able to rattle off with equal ease. Though the show was far less of a Willie Howard vehicle than *Sky High*, the comedian did get the chance to do his imitations. They were worked into the script when the momentarily unhappy heroine sings the forlorn "But Not for Me," and the comedian, to cheer her up, reprises the ballad in the styles of Chevalier, Jolson, Cantor, and Jessel.

If *Girl Crazy* was easily Willie Howard's most celebrated book musical, the 1931 *George White's Scandals*, in which Eugene joined him, was surely among his most celebrated revues. The show's score by Lew Brown and Ray

*Note that in this lyric and in the one for "Bidin' My Time," also in *Girl Crazy*, the stressed rhyme falls on the same two words, "time" and "I'm."

Henderson (partner B. G. DeSylva had left to become a Hollywood producer) yielded such winners as "Life Is Just a Bowl of Cherries," "The Thrill Is Gone," "That's Why Darkies Were Born," and "My Song," which were sung by *Girl Crazy* alumna Ethel Merman (who joined the cast just before the show reached New York), Rudy Vallee, and Everett Marshall. Ray Bolger provided the dancing.

In their opening scene, the Howard brothers, as celebrities being interviewed by a newspaper reporter, indulged in their ancient form of setup-and-gag dialogue. Sample:

WILLIE: I bagged a lion in Africa.
REPORTER: You bagged a lion?
WILLIE: I bagged him and I bagged him but he vouldn't go avay.

Things picked up after that. In a sketch set in Venice, Willie was the French ambassador romancing a girl in a gondola whose husband, played by Eugene, was the easily bribed, less-than-masculine gondolier. In "The Fleischmann Hour," Willie was Dr. Bolgareen of Budapest, a celebrated chemist ("You have got to excuse mine English as I am still very short in America"), who gives a garbled endorsement for Fleischmann's Yeast, Vallee's radio sponsor. In "The Duel," Willie was an amorous general involved with a beautiful spy. In "The Daily Reflector," Willie was the editor of a scandal sheet who shoots his ace reporter, Walter Windshield (Ray Bolger), because the paper has already gone to press with a front-page story that Windshield has been rubbed out by gangsters. And in the final scene, not only did Willie get to do his impression of Al Jolson (singing "Life Is Just a Bowl of Cherries"), Vallee also got to do *his* impression of Maurice Chevalier.

Unquestionably, the standout sketch was "Pay the Two Dollars" by Billy K. Wells, in which Willie appeared with Eugene. In the Kafka-esque episode, Max Pincus (Willie) and his lawyer, Abe Steiner (Eugene), have an argument on a subway train, and Max spits on the floor. He is willing to pay the two-dollar fine, but Abe insists on fighting the matter in court. He goes through appeal after appeal, totally ignoring Max's pathetic wail, "Pay the two dollars," until it is discovered that the spitting caused germs to spread and Max is sentenced to be hanged. At the last minute, however, he is pardoned by the governor. Back on the subway train with his lawyer, Max wails, "You spent all my money, you ruined my business, my wife left me. You're a lawyer? Pooh!" And again he spits on the floor . . .*

*When this sketch was filmed as part of the 1946 movie revue, *Ziegfeld Follies*, the roles were played by Victor Moore and Edward Arnold.

Willie's reviews were among the most favorable he ever received. "Willie Howard is more excitably funny than ever," wrote Brooks Atkinson in the *Times*. "He screams, saws the air with his sharp fingers, rushes in alarm around the stage and intersperses with some of the blandest gleams ever shot across a footlight." According to the *Post*'s John Mason Brown, "He is unflaggingly funny, so funny in fact that he enters the ranks of our foremost comedians."

Ballyhoo of 1932 was a disappointment, notwithstanding the presence of the Howard brothers and such other comic performers as Bob Hope, Paul Hartman, and Lulu McConnell. Inspired by the popular risqué magazine called *Ballyhoo*, it offered, as something of a theme, the notion that ballyhoo was the cure for the nation's economic woes. For this revue, Willie not only brought back his *Rigoletto* Quartet burlesque but added a bit of business that made it even more popular than before. On opening night, clad in formal attire, as were the three other vocalists, he began as usual by exaggerating the grimaces, gestures, and vocal flourishes of a serious, dedicated concert singer. Soon, however, Willie became distracted by the heaving bosom of the soprano in a low-cut gown standing beside him and, impulsively, he began sneaking glances in her direction. As the singing continued, he found it increasingly hard to maintain his furrow-browed concentration on the music, and his adolescent glee became more and more uninhibited. The routine was so well received that Willie repeated it—with or without variations—in four subsequent Broadway shows, plus the 1935 Paramount film, *Millions in the Air*.

It was in *Ballyhoo* that Willie also unveiled his monologue, "Comes the Revolution," in which he was seen as a Columbus Circle radical. Mounting a soapbox, he spots three passersby. "Hey, you dere," he calls. "Here, make a soycle, make a soycle. Dot's enough. Don't crowd. Don't crowd." Though he has only a straggly audience, he proceeds with his speech: "Fellow voykehs, de time has arrived. Our cop of beeterness eet ees feeled to de breem. Ve must t'row off de yoke of oppression, de capitaleests . . . Rewolt! Rewolt! I'm callink on all de voykink classes to rewolt!" At the end of his exhortation, he specifies some of the capitalistic privileges the ruling classes will no longer enjoy: "Comes de revolution, no more seelks and satins. Comes de revolution, no more limozeenes. Comes de revolution, ve'll eat strawberries and cream!" When one of his listeners says he doesn't like strawberries and cream, Willie knows just how to handle a troublemaker like that. "Comes de revolution," he threatens, "you'll eat strawberries and cream and like it!"

Following their three-month stay in *Ballyhoo*, the Howard brothers joined the cast of the then-running *George White's Music Hall Varieties*, a show more vaudeville than revue, which also featured Harry Richman, Bert Lahr, Tom Patricola, and Eleanor Powell. As an economy move, White did away with elaborate sets, costumes, and production numbers, and chose most of the material from past editions of his *Scandals.* The top ticket price was $2.50. The Howards were seen in three sketches: "The Feud" from the *Scandals of 1926*, and "The Duel" and "Pay the Two Dollars" from the *Scandals of 1931*.

Willie and Eugene's only appearance in a *Ziegfeld Follies* occurred after the showman had died and his longtime rivals, the Shubert brothers—with Ziegfeld's widow, Billie Burke, fronting as titular producer—brought the revue back to Broadway in 1934. Here the boys were co-featured with Fanny Brice, with dancers Buddy and Vilma Ebsen, ballerina Patricia Bowman, singers Jane Froman and Everett Marshall, and comedienne Eve Arden also in the cast. At least two of Willie's routines had had their origins in previous revues. As a member of the sober-faced Follies Chorale Ensemble he joined in spoofing "Who's Afraid of the Big Bad Wolf?" as if it were grand opera, which, of course, still found him ogling the mammae of the girl next to him. And in "All Quiet in Havana," author H. I. Phillips was inspired by the sketch about the Mexican president in *George White's Scandals of 1929*. In the new one, Willie, with his Yiddish-Spanish accent, portrayed both the outgoing and the incoming presidents of Cuba. Clad in a huge sombrero that found him staggering under its weight, Willie's newly elected Gonzales Machado enters to proclaim the New Raw Deal, then turns around to reveal a bull's eye painted on his back as the best protection against assassination.

Willie appeared with Fanny Brice in a take-off on the popular Broadway play, *Sailor, Beware!*, called "Sailor, Behave!" The David Freedman skit was one of the rare occasions that took advantage of the comic possibilities inherent in depicting Willie not as a schnook or a scamp but as a glib ladies' man. The comedian played Dynamite Moe, "The Gable of the Gobs," who has made a bet with his fellow sailors that he can make out with Upright Annie, the man-resisting belle of Panama. Most of the scene consisted of Willie unsuccessfully chasing Fanny around a large bed.

It was not until the twelfth edition of *George White's Scandals* was playing in Washington in 1935, two weeks before its Christmas opening on Broadway, that the producer sent an urgent appeal to Willie and Eugene—as well as to Rudy Vallee—to help rescue the floundering show. Ever since its disastrous tryout opening in Richmond, the revue had been experiencing a number of crises (including the loss of singer-comedienne Lyda Roberti), but White was determined that he could still turn it into a hit. Since the show already had one major comic, Bert Lahr, the addition of the Howards was indication of the producer's reliance on laughmakers to help pull the show through. The problem was that except for their dependable *Rigoletto* Quartet burlesque, the brothers had nothing ready and new material had to be hastily written. Among the additions: a client-lawyer sketch (Willie explains to Eugene that he wants to kill his wife because it's cheaper than alimony), and a new character for Willie, Professor Pierre Ginsberg. In the monologue, the supremely self-assured teacher—complete with tousled hair and elegantly waxed mustache and goatee—attempts to give a French lesson over the radio but only succeeds in thoroughly confusing the language with Yiddish. The routine soon became as much of a trademark as Willie's *Rigoletto* Quartet or his imitations.

The 1935 *Scandals* was not among Willie's most memorable revues, but it did prevent him from taking the co-starring role opposite Ethel Merman in the new Cole Porter musical, *Red, Hot and Blue!* Because he was then touring in the George White revue and was bound by a run-of-the-play contract, Willie was unable to accept the offer to replace Victor Moore who had pulled out of *Red, Hot and Blue!* before rehearsals. Eventually, the part went to Jimmy Durante.

Willie Howard, however, did get to succeed another comedian his next time on Broadway. Vincente Minnelli's revue, *The Show Is On,* which the Shuberts had closed in August 1937, was back in September at the Winter Garden for a brief pre-tour engagement. While vaudevillian Rose King was considered no match for the show's original female star, Beatrice Lillie, Willie was deemed at least as worthy as its original male star, Bert Lahr, and he skillfully adjusted the material to his own measurements. This was especially noted in his version of "Song of the Woodman," the Harold Arlen–E. Y. Harburg number that had been considered almost non-transferable Lahr property. Yet Willie, with his long experience in deflating pretentious concert singers, was well equipped to handle the role of the outdoorsy baritone, even adding an extra note of ludicrous pathos to the moment when wood chips descend all over his head. The only specialty routine that Willie brought with him to *The Show Is On* was his by-now celebrated appearance as Professor Pierre Ginsberg.

In 1939 it was back to the *George White's Scandals* for Willie and Eugene, their sixth edition and the final one in the series. Because of the influence of *Hellzapoppin,* the long-running Olsen and Johnson madhouse revue, the emphasis was more on slapstick humor than in previous editions, thus accounting for the presence of Ben Blue and the Three Stooges. The singing was taken care of mostly by Ella Logan and the dancing by Ann Miller (her first time on Broadway). The sketches were not overburdened with subtlety. In one scene, Willie played a French poodle on all fours "saluting" a Nazi dachshund. In another, he was a professional labor agitator whose lovemaking with his wife is interrupted by John L. Lewis (played by Eugene) because the lady refuses to sign with the union. In another, reminiscent of "Pay the Two Dollars," Willie is conned into buying a $4.85 radio wholesale, but by the time the deal is over he has paid fast-talking Eugene ten times the amount. In "Tel-U-Vision," Willie is able to drop a quarter into a machine which allows him to keep tabs on his wife while he is away.

Hard to believe, the first-act finale was yet another "Tin Pan Alley" panegyric, with music teacher Ella Logan attacking popular music, and Willie Howard, Ray Middleton, and others defending it by offering such samples as "My Old Kentucky Home," "Ol' Man River," and "Three Little Fishes." The number concluded with a tribute to George Gershwin.

After Eugene retired in 1940 (he died in 1965 at the age of eighty-four), Willie never again enjoyed success in a Broadway musical comedy or revue. He appeared the following year in *Crazy with the Heat* opposite deadpan comedienne, Luella Gear. Originally produced and directed by Kurt Kasznar (later to become a familiar face in Hollywood movies), the revue was withdrawn after only one week because of uniformly negative reviews. It was then taken over by columnist Ed Sullivan, who had it restaged (by Lew Brown) and augmented by new singers and dancers. It still lasted no more than three months.

Willie's material was considered unworthy of him, though he did manage to garner laughs when he appeared as an aging Russian ballet dancer whose right hand clutches a rose and whose spindly legs are encased in a pair of wrinkled tights. He also found some merriment as a Russian singer in a restaurant who succeeds only in annoying the customers, and as a thoroughly baffled "Voice of Experience" on the radio who must try to answer the problems of double-talking Harold Gary.

Early in 1942, night-club impresario Clifford C. Fischer decided that what wartime Broadway most needed was good old-fashioned vaudeville. Calling his show *Priorities of 1942*, Fischer offered audiences a stellar lineup (Willie Howard, Lou Holtz, Phil Baker, Paul Draper, and Hazel Scott), twelve performances a week, and a top price of $2.20. Despite mixed reviews, the show caught on, ran 353 performances, and spawned about a dozen more such entertainments, though none as successful. Willie's material consisted only of routines he had performed the previous year in *Crazy with the Heat*, such as "Voice of Experience," this time with double-talking Al Kelly.

Willie returned to more—or less—traditional fare the following year, though *My Dear Public*, which was subheaded, almost apologetically, "A Revusical Story," was simply a book musical without a book. In it, the comedian played a zipper magnate whose wife gets him to invest in a musical, but *My Dear Public* soon disintegrated into another vaudeville show. The "Comes the Revolution" bit was dusted off, and Willie got further mileage with his look of perplexity mixed with disgust as he tried to comprehend Al Kelly's double-talk. In a variation on the *Rigoletto* spoof, he remarked with justifiable concern, "What I really need is my brother Gene and two more girls."

Broadway did not see Willie Howard until four and a half years later. In the meantime, he appeared in the Chicago company of Michael Todd's *Star and Garter*, taking over the songs and routines that had once been Bobby

Clark's (including his own version of "I'm Willie the Roué from Reading, PA."). The show was followed by the ill-fated attempt to revive *The Passing Show*, which also closed in Chicago.

Willie's return to Broadway was prompted by a brief vogue in the mid-1940s of outfitting ancient musicals with the antics of popular stage comedians. It started in 1945 when Eddie Foy, Jr., appeared in a successful revival of Victor Herbert's *The Red Mill*, and continued in 1947 with Bobby Clark in Herbert's *Sweethearts*, which did almost as well. Willie's turn came the following year when Hunt Stromberg, Jr., signed the comedian to star in a new production of the 1920 Jerome Kern musical, *Sally*, with Bambi Linn as the heroine. This time the formula didn't work, despite Willie's immensely comic performance as the Grand Duke Constantine who must earn his bread working as a waiter in a Greenwich Village restaurant. The highlight of his performance came when wearing a messy wig, a droopy mustache, and a baggy suit, he delivered the sentimental ballad, "Tulip Time in Sing-Sing" ("How I miss the peace and quiet/ And the simple, wholesome diet"), which had first been heard in the 1924 show, *Sitting Pretty*. As he sang, Willie transformed himself into the total concert-hall singer, pursing his lips for the proper pear-shaped tones and closing and popping his eyes as the ardent melody engulfs him. Since *Sally* was little more than a vehicle, Willie made sure to bring along some familiar routines: his imitations of Cantor, Jessel, Jolson, and Chevalier were forced into the plot as the means through which he illustrated to stage-struck Sally just how to put across a song; the *Rigoletto* romp and "The French Professor" were added a week after the show opened. But nothing could help and *Sally* closed in a month.

On June 5, 1948, Willie made his final appearance in *Sally* and also his final appearance on the Broadway stage.

Willie Howard never enjoyed the kind of legendary status bestowed on the other clowns discussed in this book. No list of great funnymen is likely to put his name at the top. Yet as much as anyone discussed in these pages, he was a consummate stage comedian, totally dedicated to the art of making people laugh. Within his range, there was no one more masterful at timing, vocal inflection, and characterization. Though never credited as the author of his revue sketches, he was an integral part of their creation, elaborating and developing all the comic possibilities inherent in the basic situations and lines.

"The success or failure of a scene," Howard once said, "is dependent on what tricks a comedian can bring to it. First he must believe in the scene. Intuition tells him more than audience appreciation. His next step is to analyze not only the scene but the construction. To succeed, comedy must be tricked. Mugging, gestures, tempo, and vocal pyrotechnics play important roles. When you cannot get over with a line, you must trick with an absurd facial expression or a burlesque movement of some part of the anatomy, or by a surprise pitch of the voice either up or down the register. The veteran stage comic knows these tricks and uses them every second of the time he is facing an audience. Not for an instant does he relax from the broad caricature he has drawn. If he does, the illusion is gone."

Bert Lahr

ONE OF THE THEATRE'S most cherished articles of faith is that inside every great clown lurks a great dramatic actor anxious to get out. Somehow, that a comedian has mounted to the top of the exacting, disciplined, and difficult art of making people laugh is not considered truly fulfilling: He must also secretly yearn to make people cry. There is, of course, something condescending in this view since a dramatic actor is seldom thought to harbor any desire to be a clown. If he or she is ever suspected of such inclination it is considered—and hoped to be—but a temporary aberration.

This assumption about our theatre comedians is not only largely untrue it is also basically unrealistic. A comic learns early on what he can do best on the stage, and he usually spends the rest of his career perfecting his art. His bond with the audience owes much to what that audience expects of him; his characterization may be altered but it is still fundamentally the same familiar personality that has already won a following. Thus the rapport between performer and public is established every time he makes an appearance. Knowing what kind of comedy to expect, the audience is comfortable in his presence; knowing that the audience knows what to expect makes the comedian not merely comfortable but confident enough to experiment. Obviously, the clowns who have been responsible for—and revel in—the joyous sound of laughter would prefer to improve, change, and redefine an established characterization rather than risk possible failure in a different area of the theatre.

It is ironic that, except for Zero Mostel, the buffoon most often cited as having triumphed in both theatrical worlds of comedy and drama is Bert Lahr. Though every performer has a good deal of insecurity and self-doubt, there is common agreement that Lahr achieved the unenviable feat of developing normal anxieties into a state of near-total paranoia. According to Beatrice Lillie, with whom he co-starred in *The Show Is On* and *Seven Lively Arts*, "He was the only performer I knew in my life who was jumpier than I at rehearsals.

He used to fidget with his buttons while he fussed with his lines, with never a smile on his mournful jowls. 'Yeah,' he'd say, at any gag, 'but I don't think it'll do.' " Ethel Merman, with whom he appeared in *Du Barry Was a Lady*, summed him up with these words: "I don't think I've ever worked with a more talented or insecure man. I never knew anyone else who took everything so seriously—to the point where he was his own worst enemy. During rehearsals he was convinced I had received all the funny lines. He had nothing. Now Buddy DeSylva and Herbie Fields [the co-librettists] had made it a point to divide the laugh lines equally, but nothing could convince Bert of that . . . Opening night he was hilarious. Belly laughs! Gee, he could really get them. But he convinced himself that the producer had a claque out there laughing it up for him." Miss Merman further revealed that when Lahr's agent congratulated him after the smash opening, the comedian's reaction was, "Yeah, but what do I do next year?"

Abe Burrows, who directed Lahr in *Two on the Aisle*, appraised his abilities and idiocyncrasies in his autobiography. "He was a fascinating, talented, kind, generous, cantankerous, insecure, wonderful comedian," Burrows wrote, "and I think he was the funniest man who ever worked on the legitimate stage . . . When he was saying a funny line, nobody else onstage was allowed to move. It's a fact that a quick move on the part of any other actor can kill a laugh . . . But Bert carried this edict too far. When he was doing something funny, he wanted the other actors to stand absolutely still." On one occasion, when Lahr complained to Burrows that an actor had been moving, the director told him that he had been watching the scene and that the actor had not moved at all. Lahr's reply: "You're wrong. Tonight he was moving his facial muscles."

Lahr's unremitting insecurity, his background as a knockabout comedian, and his dedication to making people laugh would hardly seem to indicate an actor who would shift gears in the latter part of his career to enter new and challenging areas of the theatre. It was not, in fact, a planned change. What found Bert Lahr in the unlikely role of the tramp, Estragon, in Samuel Beckett's existential tragic-comedy, *Waiting for Godot*, was simply the conditions of the time. Despite the accolade and success that Bobby Clark was achieving during the 1940s, Lahr was sure that the theatre clown had been overtaken by the radio quipster and the more meaningful musical theatre pioneered by Rodgers and Hammerstein. Fearful that time might pass him by, he made the adjustment that extended his career, not as a tragic actor, but as a comic actor who could also play tragedy.

Lahr once said, "If a buffoon can't make an audience cry, he isn't great." Notwithstanding the fact that such indisputably great stage clowns as Beatrice Lillie, W. C. Fields, Bobby Clark, Groucho Marx, and Ed Wynn never tried to induce as much as a single tear, Lahr's comment is important because he believed it, and because it was true of his own career. Though there was never anything sentimental about his clowning, in almost everything he did was the feeling of possible danger. Bert Lahr showed the pain of comedy more than any other comic, getting his laughs by revealing his terror prompted by a variety of physical and mental blows that he responded to with an exaggerated pugnacious intensity. Lahr's guttural "gnong, gnong, gnong," his most famous vocal trademark, was usually uttered not as a meaningless laugh-provoking sound but to register amazement or disbelief at the situation in which he found himself. Even his other closely identified line, "Some fun, eh kid?," was a sarcastic expression as he tried to buck up his spirits in the face of imminent danger.

Lahr, who looked like an unhappy Happy Hooligan, had the most natural clown face of all the great comedians. Or as Gilbert Millstein once described it, "It appears to be a face composed of ridges, crevasses, and hummocks between two vast outcroppings of ears—a veritable terminal moraine of a face." Lahr's most distinctive physical characteristics were a swollen mass of a nose and two beady blue eyes, set closely together and supported by two fleshy pouches. His mouth was large and seldom closed, bellowing and braying and emitting deep long notes in a comically exaggerated vibrato. Such an appearance was meant for double-takes, dopey incomprehension, irritated huffing and puffing, forced bravado, and an almost unavoidable expression of comic anguish. Though in his early years he had a trim physique (he stood 5′9″), with a long face topped by unruly hair, he later developed a paunch, his face became rounder, and his hair almost non-existent.

At the beginning of his Broadway career, Bert seemed to place undue emphasis on all-out mugging, and his sweatful clowning appeared forced, especially when compared with that of the more spontaneous Jimmy Durante. But Lahr soon learned the value of control and the importance of letting the comedy evolve naturally from the situation and character. In the 1930s, in fact, he began to reveal a more satirical side in which he rubbed the gloss off the high-born and the pretentious, two targets he continued to take aim at throughout most of his Broadway appearances.

Bert Lahr, né Irving Lahrheim, was born in Yorkville, New York's German neighborhood on the east side of Manhattan, on August 13, 1895. His immigrant father was a hard-working though impecunious upholsterer, who had little concern or affection for his wife or his son. With hardly any communication at home, Bert found that theatrical entertainment provided a happy escape from the grim, silent atmosphere of the Lahrheim household. His performance in a school production was the first time he ever received recogni-

A scene from Bert Lahr's first Broadway hit, *Hold Everything!* (1928), in which Victor Moore was also featured.*

Bert Lahr broadcasting after winning the big race in the 1930 musical, *Flying High*. Others in the scene are Grace Brinkley, Henry Whittemore, and Kate Smith.

A handbill, with caricatures by Alex Gard, for the 1936 revue, *The Show Is On*. (Note that Vincente Minnelli's name is misspelled.)

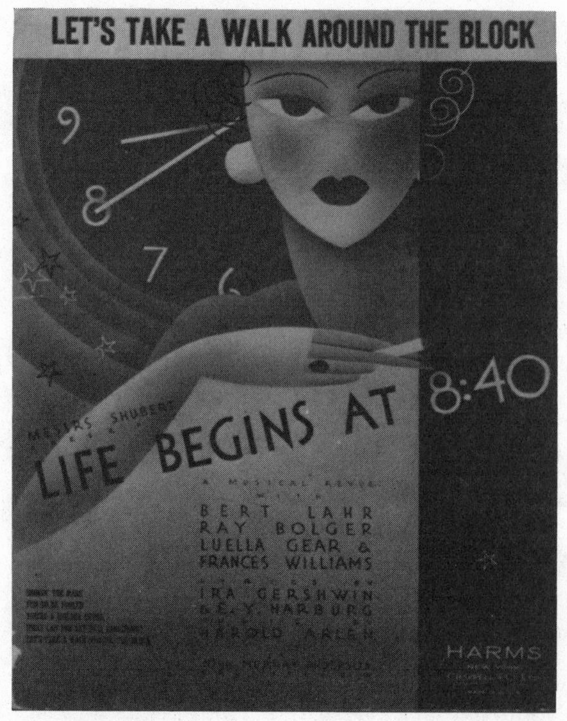

The sheet music cover of the 1934 revue, *Life Begins at 8:40*.

Ethel Merman as DuBarry and Bert Lahr as Louis XV in *DuBarry Was a Lady* (1939).

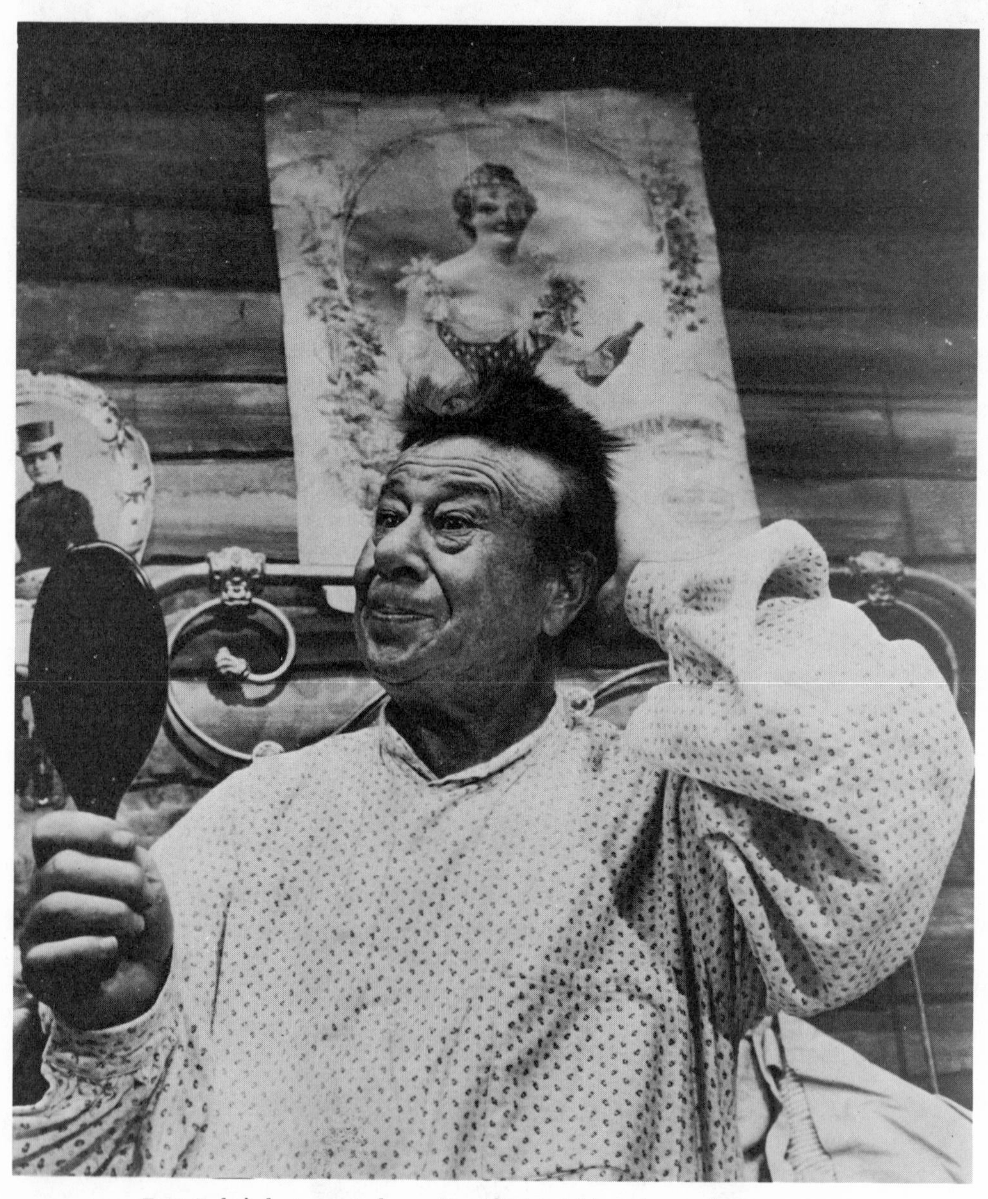

Bert Lahr's last musical was David Merrick's 1964 production, *Foxy*.

tion, and it gave him a feeling of fulfillment that—throughout his life—he could experience only in performing on the stage.

An indifferent student, Bert quit school after failing to pass the eighth grade. Following a multitude of jobs, he was invited by a friend to join a small-time vaudeville act, and from then on he knew no life other than that of the theatre. At first Bert's comic personality was based on the antics of German dialect comedians—known as Dutch comics—with an accent copied from people in his native Yorkville. Whether suitable or not, he even spoke with a German accent in a series of "school" acts, largely dealing with the outrageous behavior of youngsters in a classroom. At the age of nineteen, now too old to play a kid part, he tried breaking in an adult act in vaudeville but with little success. It did succeed, however, in attracting the attention of William (Billy) K. Wells, a writer for the high-quality burlesque chain known as the Columbia Burlesque Wheel. Wells, who would later write sketches for both the *George White's Scandals* and the *Earl Carroll Vanities,* was so impressed with Lahr's gifts that he arranged to get him a contract at Columbia. At that time, burlesque differed from vaudeville principally in that it offered companies performing in loosely constructed, fun-poking, rowdy "book" musicals rather than variety acts. It was so far removed from the age of the strip tease that chorus girls were required to wear tights.

In burlesque, Bert wore heavy makeup to give him a mature appearance, a putty nose, and a fake mustache, and he continued to use his German accent. This experience proved invaluable in developing a rapport with his audience and in helping him to cope with the competition of other comics in the same show. As his son, John Lahr, wrote in his biography of his father, *Notes on a Cowardly Lion,* "On stage, Lahr was the most frivolous of all, the loudest howler, the wildest acrobat, the merriest Merry Andrew." For five happy years, during which time he was elevated to the position of top banana, or number-one comic, he scampered, snorted, bellowed, growled, tumbled, ogled, and grimaced his way from one end of the country to the other.

Lahr's first wife, Mercedes Delpino, who had been a dancer in the Columbia Burlesque circuit, became his partner in 1922 when Bert took his next inevitable step up to big-time vaudeville. The team, billed as Lahr and Mercedes, traveled on the Orpheum Circuit for about five years. During all this time, Lahr and his wife performed only one act. Called "What's the Idea?," the sketch was something that Bert had thought up and then co-authored with Billy K. Wells. The reason he never changed the act was that it became so popular theatre managers and audiences simply wouldn't let him. Based on the classic low-comedy situation contrasting a sexy beauty with a klutzy beast, the skit takes place on a street corner and opens with Mercedes in a sequined dress singing a seductive song followed by a bump-and-grind dance. Suddenly, a loud banging noise is heard, and Bert, wearing a baggy policeman's uniform with his cap twisted to one side, comes bounding downstage, beating his night stick on the street and bellowing in his "Dutch" accent, "Shtop, in the name of the shtationhouse! Shtopppppppppp! What's the idea? What's the idea? What's the idea of massagin' the atmosphere?" Mercedes's gyrations, it seems, are a violation of the law, which Bert explains is the "Nineteenth Amendment, Section Six, Upper Seven, which says it's a public nonsense to shimmy or vibrate any part of the human astronomy. And it's punishable by the fine of one year or imprisonment for two years of E Pluribus Aluminum."

"But what was I doing?" Mercedes pleads. "I wasn't doing anything wrong." "You wasn't doin' anything wrong?" screams Bert. "Gnong, gnong, gnong! I saw ya. I was standin' right down there, and the second I saw ya I said to myself, 'This has got to stop at once!' So I watched ya for ten minutes."

Bert attempts to take Mercedes to the station house, but she pushes him away, knocking his cap over his eyes. Lahr, now realizing he is up against a tough adversary, walks around her in a crouch, snarling and snorting his authority while brandishing his night stick in her face. But he loses his dignity again when his club somehow gets caught behind his left leg, and he trips over it and falls on his face. Now Mercedes threatens to report Bert for drunkenness, but he soon discovers that she is his niece ("I didn't see ya since your infantry. My goodness, how time flitters."). This somehow prompts Bert to render the song, "Peggy O'Neill," and Mercedes offers "La Soldata," with Bert parodying her movements. When he still insists on taking Mercedes to the station house, she begins to vamp him, and Bert dreamily succumbs to her advances. His pleasure is abruptly halted when he sees his police sergeant walking down the street and he shoos Mercedes away. Greeting the sergeant, Lahr feigns coolness by flipping his night stick, which flies out of his hand. When the sergeant wants to know what he was doing with his arm around the girl, Bert has a perfect excuse: "I was friskin' her. She's a very dangerous character." "Why didn't you pinch her?" asks the sergeant. "I did," admits Lahr, who quickly catches himself. "I mean I gave her a ticket."

As the two men talk about the upcoming policeman's ball the scene ends somewhat incongruously with the appearance of a jealous French girl, who, coming upon her former lover and his current flame, shoots them and then commits suicide. But Lahr and the sergeant, totally impervious to the mayhem, continue talking about the ball as they stroll offstage.

When, in 1925, Lahr and Mercedes were headliners at the Palace, it was clear that vaudeville would someday have to share them with Broadway. Two years later, Harry Delmar, a vaudeville tap dancer with Ziegfeldian aspirations, began assembling material for a revue, *Harry Delmar's Revels.* With Billy K. Wells writing the sketches, Delmar signed the stellar lineup of Frank Fay, Winnie Lightner, and the team of Blossom Seeley and Benny Fields (who dropped out before the New York opening). Since Fay was a slow-talking raconteur, Delmar also engaged Bert Lahr as something of a comic contrast, and along with Bert came "E. Mercedes" (as she was billed).

Harry Delmar's Revels ("Broadway's Newest and Most Spectacular Revue") opened at the Shubert Theatre on November 28, 1927. Lahr and Mercedes performed "What's the Idea?" under the new title, "Limbs of the Law," and Bert appeared in three other sketches (in one, "Four Famous Horsemen," he was seen as Jesse James). The comedian also had been slated to sing the Jimmy McHugh–Dorothy Fields "I Can't Give You Anything But Love" as a duet with Patsy Kelly. In the scene, which was to have taken place in front of a jewelry store, Bert and Patsy were two raggedy kids confessing that they had nothing to give each other but love. But Delmar, who had planned to have the scene then turn into a dream sequence with show girls parading as various kinds of jewelry, was dissatisfied with both the song and the performance, and he scrapped the number before the opening. The *Revels* received mixed notices and, when it failed to attract an audience, the cast bought out the producer and kept the show going for three months. Changes were constantly being made, including the restoration of "I Can't Give You Anything But Love" at the end of January, but it was now sung as a solo by Lew Mann. (Three months later, the song received the attention it deserved when it was reintroduced in *Blackbirds of 1928.*) Not long after the *Revels* had begun, Winnie Lightner was replaced by Lillian Roth, and Mercedes, who was just beginning to show signs of the mental illness that would destroy her marriage to Bert, also left the cast.

Lahr's well-received antics did not go unnoticed by Broadway producers. In fact, each of the comedian's first three book musicals found Lahr appearing under the sponsorship of some of the most prestigious names in the theatre: Alex Aarons and Vinton Freedley *(Hold Everything!)*, George White *(Flying High)*, and Florenz Ziegfeld *(Hot-Cha!)*. While the production staffs were different in all three, the shows were cut largely from the same pattern. All had music by Ray Henderson and lyrics by Lew Brown, with Buddy DeSylva as co-lyricist of the first two. DeSylva was also the co-librettist with John McGowan of *Hold Everything!* and *Flying High*, and Brown and Henderson shared book credit with Mark Hellinger for *Hot-Cha!* The three swiftly paced shows were concerned with farcical stories dealing with professions that require not only skill but daring, and in all of them Bert had to undergo elaborate, side-splitting preparations for a risky task from which he somehow emerged triumphant. Although he had long since abandoned his sauerkraut accent, Bert played characters named, consecutively, Gink Schiner, Rusty Krause, and Alky Schmidt.

In 1928 Aarons and Freedley, who had already presented book musicals starring Fred and Adele Astaire and Gertrude Lawrence, signed Bert to appear as Gink Schiner in *Hold Everything!* and gave him a five-year contract. Despite a title that suggested the show was about wrestling, *Hold Everything!* was a saga about the pugilistic wars, with Bert as a punch-drunk pug (when asked in one scene what he was reading, Gink replies, " 'Da Woiks' by William Shakespeare"). Gink is the stable-mate of welterweight contender Sunny Jim Brooks, played by Jack Whiting. Sunny Jim, it must be noted, has had to become a professional boxer to earn enough money to pay his college tuition. By the time the show is over, of course, he has won both the welterweight crown and faithful Sue Burke (Ona Munson).

Bert Lahr was not the only major comedian in *Hold Everything!* Contrasting with his rowdy personality was the bumbling, wistful Victor Moore, playing Nosey Bartlett, the training-camp cook and Gink Schiner's best friend. (Though Moore's career as a Broadway leading man dated back to 1906, he would not achieve theatrical immortality until three years after *Hold Everything!* when he appeared as Alexander Throttlebottom in *Of Thee I Sing.*)

Lahr's two big comedy scenes established the comedian among the front ranks of Broadway clowns. In Act I, which takes place in the training-camp gym, Gink's trainer has insisted that he lose three pounds and, much against his will ("Ten minutes in there and I'll be among your souvenirs"), the fighter enters a huge cabinet called an electric reducer. After Nosey switches on a light current, Gink's girl friend, Toots Breen (Nina Olivette), pushes the lever to increase the voltage just to show everyone what a he-man her Ginky is. Suddenly, Gink's face, which sticks out above the cabinet, is twisted into a grimace of agony, and he is so overcome by the increased current that he cannot utter a sound when he tries calling for help. Nosey pushes the lever back, but Toots again increases the voltage. This time, however, the lever gets stuck and Gink's face registers a variety of expressions showing how excruciating the ordeal is. His head disappears inside the cabinet, and Nosey methodically reads the machine's book of instructions ("If cabinet gets out of order, ship to our factory in Milwaukee and we will replace it"). As Nosey and Toots struggle frantically to turn off the current, there is a loud explosion, the cabinet falls apart, Toots screams, and Gink tumbles out, his face totally black.*

**Whoopee*, which opened on Broadway just two months later, had a scene in which Eddie Cantor hides in an oven. It explodes when accidentally lighted, and Eddie is thrown out with a blackened face.

Act II's comedy highlight takes place in the dressing room at Madison Square Garden on the night of the big fight. Billed as "The Waterfront Terror," Gink shadow boxes around the room with an elaborate show of self-deceptive confidence. Nosey, however, is none too impressed with his footwork, his timing, or his defense.

NOSEY: Why do you wear those pads for?

GINK: 'Cause every time I fight my knees get scraped.

NOSEY: Look out, you'll foul yourself.

GINK: Leave me alone. I'm winning. *(He stops shadow boxing and sits down looking depressed.)* Nosey, something tells me this fight is gonna be the turning point of my career.

NOSEY: What are you feelin' bad for?

GINK *(shamefully)*: Well, I bet on Kid Fracus against myself.

NOSEY *(relieved)*: Don't give it a second thought.

GINK: But I'm afraid I'm gonna win!

NOSEY: Don't let that worry you. I know you'll be good to the last drop.

Gink's premonition turns out to be correct, and he somehow wins the match (which is fought offstage). Returning to his dressing room in unexpected glory, he preens and postures in a manner befitting a victorious gladiator. Obviously, this is a role he has never played before. Feinting and jabbing the air, his eyes ablaze with exultation, Gink shouts to everyone with uncontrollable amazement and pride: "DID YOU SEE ME? DID YOU SEE ME?"

The critics readily acknowledged, as the headline of the review in the *American* awkwardly put it, "NEW COMEDY KING CROWNED IN MUSIC PLAY IN BERT LAHR." Brooks Atkinson in the *Times* maintained that "he dances, grimaces, rolls his eyes and sings with the sort of broad abandon that is instantaneously appreciated in the abdomen." Percy Hammond in the *Herald Tribune* that "he bursts into comical extravagances that relieve monotony and make you laugh." Gilbert W. Gabriel in the *Sun* that his "facefuls of India rubber anxiety will cause corruption of your funny bone." St. John Ervine in the *World* that he is "one of the quaintest drolls I have ever seen." Robert Littell in the *Post* that he "alone is enough to make an only slightly better than average show seem much better than average." Littell went on to observe, "No one who didn't have his face and leather lungs and sudden explosions of weird sounds could make what he does seem funny. It's the humor of complete, almost manic lack of restraint."

Hold Everything ran a full year on Broadway, thanks largely to Bert's newly won popularity and a bright, catchy score that included "You're the Cream in My Coffee" and "Don't Hold Everything." Other producers were soon courting the comedian, with the most attractive offer coming from George White. Of course, there was the little matter of that five-year contract with Aarons and Freedley, who were then about to start putting *Girl Crazy* into production with Bert in the main comic part. White sued to get him out of his contract, won the case, and Willie Howard was then rushed into the role. The public, of course, knew little about this backstage wrangling. As far as theatregoers were concerned, *Flying High*, the George White musical, was accepted as an equally mirthful follow-up to *Hold Everything*, with another winning DeSylva–Brown–Henderson score (including "Thank Your Father"), and a role for Bert Lahr that again found him facing mortal danger with quaking shoulders and trembling knees.*

Flying High, which opened in March 1930, cast Bert as Rusty Krause, an airplane mechanic, who signs up as a contestant in a transcontinental airplane race to avoid marrying a mail-order bride, Pansy Sparks (Kate Smith in her second Broadway role). In a scene obviously inspired by the weight-reducing episode in *Hold Everything*, Rusty has a medical examination to test his fitness to fly a plane. First, the doctor asks some questions.

DOCTOR: What's your name?
RUSTY: Emil Krause.
DOCTOR: Born?
RUSTY: What do you think?
DOCTOR *(sharply): Where* were you born?
RUSTY: I only know from heresay.
DOCTOR *(irritated): Where were you born?*
RUSTY: Hoboken.
DOCTOR: Nationality?
RUSTY: Scotch by absorption.

The doctor proceeds to test Rusty's blood pressure and take his temperature, then taps his chest several times and listens to his heartbeat by putting his ear to Rusty's chest. He repeats this twice, and Rusty, confused by what he takes to be the doctor's romantic advances, grabs him and kisses him on the forehead. When the doctor tells the would-be aviator that he must test him for astigmatism, Rusty indignantly protests, "Look here, sir, I want you to know that I've led a clean life." Wheeling in a huge drum, the doctor ex-

*John McGowan, the co-librettist of both *Girl Crazy* and *Flying High*, had written the part of Gieber Goldfarb in the former expressly for Bert Lahr. In an early draft of the libretto, the character attempted to pass himself off as an Indian by answering a real Indian with "Gnong, gnong, gnong."

plains that he will now give the tailspin test for dizziness. Rusty enters the drum and is whirled around several times. When the drum stops, he sticks his head out and the doctor asks him to identify a pencil he holds in his hand. Rusty blinks his rotating eyes and says it's a picket fence. The test is repeated, and Rusty emerges unsteadily from the drum with his hair mussed and his eyes bleary. As he staggers about, the doctor hands him a glass container and orders, "Here, you know what to do with that." Rusty looks at the glass uncomprehendingly and then at the doctor, who has turned his back and walked back to his desk. What *is* he supposed to do with the glass? Then it dawns on him. Remembering that he has a flask of whiskey in his pocket, he takes it out and carefully pours some of it into the container. Holding it as steadily as he can, Rusty stumbles over to the doctor and taps him on the shoulder. "Here you are," he says, offering the glass. "That's all I can spare."

Rusty not only manages to get an airplane aloft for the transcontinental race, he even manages to set the world record for time in the air because he doesn't have the foggiest idea how to get the plane down. When at last he returns to earth, he is rushed to a microphone for an interview. Though Rusty can barely keep his eyes open, much less talk coherently, he is asked by the interviewer to share his future plans with his listeners:

RUSTY: Ladies and gentlemen, the next thing I'll do is to discover the North Pole.

HECKLER IN AUDIENCE: That's been done.

RUSTY: I mean the South Pole.

HECKLER: That's been done, too.

RUSTY: Then I'll discover the West Pole. That's what I'll do, I'll discover the West Pole.

HECKLER: There ain't no West Pole.

RUSTY: Ladies and gentlemen, an ignorant bystander has just told me there ain't no West Pole. I'll tell you why there ain't no West Pole. Because nobody's had the guts to go out and look for it!

After a group of squealing girls come in and gush over Rusty, he ends the scene by leading them in a song-and-dance paean to himself that ends:

Who will rise and conquer men
Then become a bum again?
Mrs. Krause's blue-eyed baby boy!

Although Richard Watts, Jr., in the *Herald Tribune* felt that Lahr "substitutes hard work and furious mugging for any true comic spirit," the other critical aisle-sitters found him hard to resist. According to Gilbert W. Gabriel, now writing for the *American,* "I think him the maddest, most tireless, funniest comedian. By the sweat of his crazily corrugated brow and all the India rubber contortions of his comical mug and his grotesquely quaking shoulder blades, he wrung such laughter out of last night's audience as no clown of this season's lot could."*

Opening two years after *Flying High, Hot-Cha!* might have been expected to repeat the success of Lahr's first two book musicals. In this one, Bert appeared as Alky Schmidt, a New York night-club comic in a floor show that is booked into a Mexico City cabaret. When Alky reveals that his first name is really Alcarez and that his grandfather was a matador (which he pronounces "mackinaw"), he is persuaded to take up the profession with the assurance that all he will be expected to do is to look fierce and dance around a bull. This, of course, leads to the inevitable scene in which Bert must endure some sort of physical training in preparation for a future event that is fraught with danger. Here it takes place in the bullfighters' practice quarters where Alky learns a variety of movements that strike him as being both easy and effeminate as he twirls and gets entangled in a cape, falls on his face, and dances with his instructor.

After being kept in the dark as to what a "mackinaw" must do that will earn him the bankroll he has been promised, Alky finally learns that he must fight a raging bull named Black Demon. Quivering with fear just before he is about to enter the ring, he is told by his manager (Lynne Overman) that he has nothing to worry about. It seems that unseen by anyone, the manager has given the bull a laxative—which Alky also mistakenly takes—and, as described by onlookers since the scene is not shown, both bull and matador spend their time in the ring backing away from each other. When last seen, Alky grandly re-enters the bullpen in triumph.

Hot-Cha!—which unblushingly used the subtitle "Laid in Mexico" in the program—might have succeeded on Broadway had not the musical theatre suddenly become dominated by a new wave of adult, sophisticated productions led by *The Cat and the Fiddle, Of Thee I Sing,* and *Face the Music.* The Lahr vehicle was simply following an old-fashioned formula at a time when the stage was alive with new ideas. Though *Hot-Cha!* was produced by Florenz Ziegfeld, the master showman was then seriously ill (he died less than five months after the opening) and showed little inclination to keep it up to date either in subject matter or style. In fact, one of the songs even poked fun at recent offerings by deriding "Park Avenue librettos by children of the ghettos."

*Bert also played Rusty Krause in M-G-M's 1931 film version of *Flying High.* It was the only stage role he repeated on the screen.

Obviously, the *Hold Everything–Flying High* formula could no longer be followed. Since Bert Lahr did not feel ready to attempt a different kind of book musical, the only alternative was to adjust his style to the revue. For his next four Broadway appearances, Bert found this form of entertainment extremely helpful in trying out new characterizations and routines that worked well with his more accustomed forms of slapstick mayhem.

George White's Music Hall Varieties, which opened in November 1932, was an economy-package version of the *Scandals,* without the accustomed elaborate settings, costumes, or production numbers. The show, which was presented at the mammoth Casino Theatre (formerly the Earl Carroll), had a top price of $2.50, and a cast that, in addition to Bert Lahr, included Harry Richman, Lili Damita, and Eleanor Powell (after a month Miss Damita called it quits, and Willie and Eugene Howard and Tom Patricola were added).

Some of Bert's sketches had been seen in previous *Scandals* (including "A Close Shave," formerly known as "Lady Barber" when Willie Howard did it in the 1926 edition), but this was the first show in which Lahr revealed a subtler, more satirical aspect of his clowning. With George White's encouragement, he kidded the svelte elegance of the popular musical-comedy star, Clifton Webb, here renamed Clifton Duckfeet, by appearing in front of the traveler curtains at center stage formally dressed in a white bolero jacket, tuxedo pants, and a top hat, with a gold chain dangling from his waist. As he sang "A Bottle and a Bird" (by Irving Caesar), he parodied Webb's precise speech and mincing footwork while recounting a romantic adventure in Paris ("A rendezvous, tête-à-tête,/ For me pajamas—Alice blue,/ For her—negligée—and fetching, too"), which ended on an embarrassingly commercial note when the girl asked for fifty dollars.*

In another scene, Lahr was again in formal attire as he sang "Trees" with all the exaggerated emotion of a concert baritone while surrounded by a group of sniffing dogs. He also did an uncharacteristic blackface bit as Al Jolson yearning for his dear old "Cabin in the Cotton."

Lahr's newfound flair for satire was given even greater opportunity in the Shubert revue, *Life Begins at 8:40,* which opened late in August 1934. Ray Bolger, Luella Gear, and Frances Williams were also featured in the production, and John Murray Anderson was the director. Composer Harold Arlen and co-lyricists E. Y. Harburg and Ira Gershwin (whose score included "Let's Take a Walk Around the Block" and "Fun To Be Fooled") provided Bert with a solo called "Things" that he sang as the second of his concert-baritone spoofs. In a scene taking place at a ladies' garden party, he appeared in white tie and

*Bert Lahr was not the only clown on Broadway to do a take-off on Clifton Webb during the 1932–33 season. Shortly after the opening of *George White's Music Hall Varieties,* Bobby Clark appeared in *Walk a Little Faster* doing a Webb burlesque in a dance with Beatrice Lillie.

tails wearing an ill-fitting toupee. After introducing the piece as "a little thing of moods and fancies," he then went on to sputter a heartfelt ballad celebrating the "utter utter utter utter loveliness of things," while nonchalantly picking his teeth. At the end, his voice quavering with emotion, he boomed lustily, "You can have your smoke-pipe rings/ And your Saratoga Springs,/ But give mee, gi-i-ive meeee thiiiiiiiiings!" With that, a little old lady hit him in the face with a pie.

In another number, "Quartette Erotica," Lahr (as Balzac), Bolger (as Boccaccio), Brian Donlevy (as de Maupassant), and James MacColl (as Rabelais) lamented the fact that modern literature had become so outspoken that their once daring efforts were no longer shocking ("The dirt we used to dish up, sad to say,/ Wouldn't shock a bishop of today").

The comic sketches further extended Lahr's range. In "C'est la Vie" (by Harburg and Gershwin), Parisians Lahr and Bolger, both wearing Inverness capes over their formal clothes, are about to jump into the Seine when they discover that they are committing suicide because of the same woman. Suddenly, the woman (Luella Gear) rushes in, reveals that her heart is torn between them, and, since she has just seen Noël Coward's *Design for Living,* has decided that they can all live happily together. With that, the two overjoyed men dance and sing and then hug and kiss each other. Now realizing that they no longer have need for a woman, they dump Luella into the Seine. In "Chin Up" (by Alan Baxter), Bert, in a silk dressing gown, leads the cast in kidding the tony accent, clipped speech, honor-of-the-family moral code, and verbal understatement of the imperturbable British upper class. In the scene, Bert is about to dress for dinner when, one after another, his father takes poison because of gambling debts, his wife takes poison because she has been unfaithful, and his mother takes poison because he was born out of wedlock. All the while, as Bert stands calmly by, toasts are offered punctuated by such bromides as "Stiff upper lip" and "Chin up." But Bert too must take poison because the suicides have made him late for dinner with the duchess. His last gasp: "Needn't dress." More in keeping with the Bert Lahr of old is "A Day at the Brokers' " (by David Freedman), in which Bert is taken in by a fast-talking stock broker (Brian Donlevy) whose hot tip finds the comedian dashing madly about the stage as he goes through convulsions at every rise and fall of the volatile silver-mine stock he has just bought.

Life Begins at 8:40 was generally applauded as a fitting successor to the Shuberts' previous revue, *Ziegfeld Follies of 1934,* with Bert Lahr's newly revealed versatility and subtlety widely appreciated by the critics. As John Mason Brown observed in the *Post,* "Mr. Lahr is now in control of his energy instead of allowing it to be in control of him. He has harnessed the Niagara of his high spirits, and, refusing to be carried along by what was once the wasteful rush of their powers, he has made them work for him. He is not the single-track zany he once seemed fated to be."

Bert Lahr returned to the George White fold for his only appearance in an edition of the *George White's Scandals*. The show experienced many problems on the road (for one thing, featured singer-comedienne Lyda Roberti left after a week), and by the time it opened on Broadway on Christmas night, 1935, Willie and Eugene Howard had been added to augment the comedy and Rudy Vallee had been added as featured male singer. Bert again did his take-off on the British nobility in the sketch, "A Blessed Event" (by Billy K. Wells), in which he was seen as Lord Marlebone chatting with his friend Lord Tottingham (Hal Forde) in their London club. Marlebone's wife telephones to inform him that she is pregnant and that Tottingham is the father. Without registering any emotion, Marlebone orders another scotch and takes out his revolver. "Awfully sorry," he tells Tottingham, "and I'm sure you won't mind, but I've got to shoot you."

Bert's comic highlight in the show was his appearance as Professor Friedrich Von Kluck, the noted Viennese culinary expert, who is seen giving a radio broadcast demonstrating how to make soup. Wearing a toupee, thick glasses, and a walrus mustache, and dressed in a white surgical gown, Von Kluck bounds into a room designed as a modern kitchen, with an intern and a nurse in attendance. Addressing his radio audience as "My dear listener-inners," the nearsighted professor washes his hands in a kettle instead of a basin, then places the needed ingredients on a table. He cries as he peels an onion and drops it into the kettle. He cleans a potato by blowing on it and wiping it on his gown. Then he carefully whittles it into nothing. Next he adds spinach. His toupee falls into the kettle and he retrieves it. He puts in carrots, eggs, and peas. His toupee falls in again and he lets it stay there. After all the ingredients have gone into the kettle, he stirs it with a fly swatter. When the microphone into which he has been speaking also falls in, he continues talking into it by bending over the kettle until he fishes it out. At last everything is ready, and he pours the soup into a bowl and tastes it. To his amazement, it's delicious! Though knocked out by the shock, he continues talking as he sinks to the floor. Blackout.

After the breakthrough of *Life Begins at 8:40*, the *George White's Scandals* was almost a throwback to a more undisciplined time, and the show barely lasted one hundred performances. But exactly one year after the *Scandals* opening, Bert returned to Broadway in one of the most memorable revues of the 1930s, *The Show Is On*. His co-star in the Shubert production was the illustrious female clown, Beatrice Lillie, and the entire production was "conceived, staged and designed" by Vincente Minnelli. All the sketches but one were written by David Freedman, though, rather than have one songwriting team, there were contributions from such major talents as Vernon Duke and Ted Fetter ("Now"), Richard Rodgers and Lorenz Hart ("Rhythm"), the Gershwin brothers ("By Strauss"), and Hoagy Carmichael and Stanley Adams ("Little Old Lady").

Lahr's material was even better than what he had had in *Life Begins at 8:40*. In general, the satirical thrusts were aimed at various aspects of the world of entertainment. To put down meaningless scat singing, Bert offered a grunting, growling, and gargling item called "Woof." To give both screen royalty and British royalty a ribbing, he appeared in the skit, "Titania," as movie idol Ronald Traylor (complete with a widow's peak that came to a point midway between his eyebrows). Based on the involvement between King Edward VIII and Wallis Simpson, the sketch was concerned with the efforts of a handsome actor, at the instigation of the prime minister, to woo commoner Sybil Hutchins (Vera Allen) away from Titania's king. As befitting the theme, the dialogue was again brittle British:

RONALD: Cocktails?
SYBIL: Thenkyoo.
RONALD: Cigarette?
SYBIL: Thenkyoo.
RONALD: Metch?
SYBIL: Thenkyoo.
RONALD: Gum?
SYBIL: Thenkyoo.

Though Ronald succeeds in persuading Sybil to go to Hollywood with him, they will not be alone. The king has just signed with M-G-M.

In "Taxes! Taxes!," probably inspired by the sketch in the broker's office in *Life Begins at 8:40*, Lahr played another movie actor, Bert Larrimore, who goes to the IRS office to demand a refund of $197.56 on the taxes he paid for 1934. The tax official (Reginald Gardiner) reviews all the obviously padded deductions for entertaining and traveling (Bert's explanation: "Buses and streetcars. A dime here, a quarter there. It adds up"), and a claimed contribution of $10,000.00 to the Arizona Foundling Home for Illegitimate Indians ("I like to encourage things like that"). By the time all the false deductions have been disallowed, it turns out that Larrimore owes the government $40,000.50 for 1934. But when the official wants to take up the matter of Bert's 1935 taxes, the actor smugly tells him that he knows there is absolutely no evidence against him. Why? "Because in 1935 I didn't file any income tax."

In a parody of the play, *Tovarich*, which dealt with imperial Russian émigrés working as servants in Paris, Bert was seen as Michael, and Vera Allen was Tatiana, Republican servants in the home of Washington New Dealers, who hide a picture of "Little Father" Herbert Hoover behind one of Thomas Jefferson. The only sketch Bert Lahr and Beatrice Lillie appeared in together was "Box Office," in which Miss Lillie played a snobbish box-office treasurer of the Theatre Guild (pronounced "Geeld" by Miss Lillie) and Bert was a critic from Cain's Warehouse passing judgment on the Guild's pretentious offerings. The two stars were also in the production number, "Burlesque," with Lillie as a stripper and Lahr as a red-nosed, baggypants comic.

The one routine in the show that will forever be associated with Bert Lahr was "Song of the Woodman" by Harold Arlen and E. Y. Harburg. Following in the spirit of Lahr's previous concert-baritone take-offs, "Trees" and "Things," this one made sport of the he-man, outdoorsy type who celebrate the open road and the joys of manual labor. Bert came out swinging an axe and wearing a toupee, a checkered lumberjack shirt, jodhpur breeches, and laced leather boots. He began his tribute with an emotional paraphrase of Browning's celebrated lines from "Pippa Passes":

> The day's at the dawn, the dawn's at the morn,
> The morn's at the lawn, the lawn's on the corn.
> The corn's on the cob, all's right with the world.
> All's rah-rah-rah-rah-rah-rah-right with the world.

Bert went on to bellow ("With a song in my soul and an aa-aa-aa-axe in my hand") the way he feels as he chops and chops and chops until the sun comes up. Then—as Jimmy Durante had done in *his* song about wood in *The New Yorkers*—Lahr enumerated all the wonderful products that come from a tree: a crib, a chair, a soap box, a bat, a gat, a rolling pin, a picket fence, water buckets, telephone poles, cribbage boards, toothpicks, croquet balls, mousetraps, magazines, flag masts, brush handles, and comfort seats ("all shapes and classes for little lads and . . . little lasses"). During the rendition, whenever he simulated chopping a tree on the lines, "Heave-ho, heave-ho, heave-ho, heave-ho," wood chips would rain down on him from the wings, upsetting his toupee and almost knocking him down.

When *The Show Is On* closed in July 1937, Bert could find nothing appealing in the scripts being offered him, and he felt the time had come to explore the possibilities of a Hollywood film career. But apart from his memorable performance as the Cowardly Lion in *The Wizard of Oz*, his West Coast experience gave him little satisfaction. Even his one success did not lead to a renewal of his M-G-M contract ("What kind of parts can you give a lion?," he once asked), and he was anxious for the opportunity to return to Broadway.

When it came, the opportunity proved auspicious. B. G. De Sylva, a former lyricist and film producer, longed to establish himself as a Broadway producer with an original book musical in which, in a dream, Mae West would appear as Du Barry and Bert would be Louis XV. Since Miss West was not interested, Bert found himself co-starring with Ethel Merman in *Du Barry Was a Lady,* one of the biggest hits of the decade.

In the story, written by Herbert Fields and De Sylva, Louis Blore (Bert), who works at the posh Club Petite, where singer May Daly (Ethel) is the featured attraction, has just discovered he holds the winning Irish Sweepstakes number. Reporters rush to question him about his good fortune:

REPORTER: Mr. Blore, what did you do when you heard you'd won the Sweepstakes?

LOUIS: For a few seconds, I didn't do nothin'. Then I fell flat on my kisser.

REPORTER: Seventy-five thousand dollars is a lot of money. What are you going to do with it?

LOUIS: Spend it.

REPORTER: I know, but how?

LOUIS: Gradually.

REPORTER: What was your job here, Mr. Blore?

LOUIS: Aww . . . I was gonna quit anyway.

REPORTER: Well, what were you, a waiter?

LOUIS: I quit, I told ya.

REPORTER: Well, what did you do before you quit?

LOUIS: Listen, it's honest work. Somebody's gotta do it.

REPORTER: Do what, Mr. Blore? What did you do?

LOUIS: All right, you asked for it. I was a ladies' maid in a gents' room.

Louis is in love with May Daly, but she prefers Alex Barton (Ronald Graham), and Louis puts a Mickey Finn in Barton's drink. Somehow their glasses get switched, and Louis takes the knockout drops. From then on, most of the story takes place in his dream in which he is the French monarch and May Daly is his far-from-willing paramour. Though Bert's Louis XV enjoys the privileges of royalty, he is also something of a reformer. To show his democratic nature, he tells a courtseying cutie, "Skip the dip," and to show his concern for his country's economy, he comes up with a new relief project: "Build a bridge across the Seine lengthwise."

Du Barry Was a Lady gave Lahr two memorable scenes. The first, in the gentlemen's lavatory of the Club Petite, finds Louis Blore instructing his replacement, Charlie (Benny Baker), on the fine points of getting the best possible tip. The most important point is to make the customer feel obligated, though Louis cautions Charlie that he cannot learn a job like this overnight ("This here's a career!"). Louis demonstrates the properly dainty manner to turn on the "facet" with his little finger "perntin'," and the proper amount of bath crystals and perfume to use. He then goes through a rehearsal with Charlie in which he acts the part of a drunken customer staggering into the room. Charlie does all right in learning how to offer a towel and how to brush a suit, but then he ruins everything by crassly sticking out his hand for a tip. Shocked at this gauche behavior, Louis shows how to be primed for the tip by keeping his right arm at his side and only, with the slightest "English," indicate that a tip is expected. When Charlie asks if he should say, "Thank you, sir," Louis offers his parting advice: "Well, you can mumble it if you want to, but never act grateful!"

Lahr's second big comic scene occurred during the dream sequence when the king's son, the Dauphin (Benny Baker), shoots the monarch in the seat of the pants with a bow and arrow. Here the humor developed not only from Lahr's ability to register a variety of pained grimaces but also from the problem of how to remove the arrow. When a doctor asks if it hurts, the king is more concerned with his dignity than his discomfort: "I'm not suffering, but I can't go through life looking like a weather vane." Du Barry, however, has a practical —if anachronistic—suggestion. "Don't let it worry you," she consoles the king. "If they never get it out you can hang a flag on it on Bastille Day." Eventually, the arrow is removed—only to be quickly replaced when the Dauphin again takes aim.

Cole Porter's score for *Du Barry Was a Lady* gave Bert and Ethel an authentic showstopper in their mock-sentimental hillbilly duet, "Friendship," in which each expressed eternal devotion to the other through such pledges as "If you're ever in a jam, here I am," and "If you ever lose your teeth when you're out to dine, borrow mine." They also shared an interminable piece of single-entendre, "But in the Morning, No," which only demonstrated how even Porter could stretch a thin idea even thinner.

Du Barry Was a Lady may have been coarse, but it was sumptuous entertainment, and for a year Bert and Ethel were the reigning monarchs of Broadway. Of Lahr's performance, John Mason Brown wrote, "Mr. Lahr is a hugely funny fellow; a comic who grows more and more comic with the years. Like Miss Merman he has boundless zest. His crossed eyes, the bullfrog croakings which emerge from his pondlike mouth, his hilarious travesties of the operatic manner or the ritual of court etiquette—all these prove reliable ammunition in his comic shotgun."

When Bert Lahr asked "What do I do next year?," the night of *Du Barry*'s triumphant opening, it turned out that his long-range apprehension was fully justified. Despite the show's acclaim, no other worthwhile stage projects came his way, so once more Lahr tried Hollywood. This time he had even less success than before. Exactly five years after opening in *Du Barry Was a Lady,* he returned to New York in *Seven Lively Arts,* a revue that had the earmarks of being another runaway hit. The producer was Billy Rose, the songs were again by Cole Porter, and Beatrice Lillie was Bert's co-star. Unfortunately, though the production was lavish (Norman Bel Geddes designed the scenery), the material was weak and the show remained only 183 performances. Then, too, with *Oklahoma!, Bloomer Girl,* and *Song of Norway* as the major Broadway attractions, wartime audiences may have reacted against a revue that seemed a deliberately outdated throwback to the Ziegfeldian extravagance of the past.

Initially, *Seven Lively Arts* was to have had a story line dealing with seven young hopefuls who come to New York planning to become a tap dancer, a painter, a ballerina, a singer, a playwright, a stage actress, and a film actress, with the plot showing how they achieved their ambitions. But except for a brief interlude at the beginning of the show, by the time it reached New York this notion was scrapped in favor of a running commentary, delivered by "Doc" Rockwell and written by Ben Hecht, which introduced scenes representing movies, opera, ballet, jazz, theatre, concert, and radio. Giving life to these "arts" were such diverse artists as jazz clarinetist Benny Goodman (leading a quintet), ballet dancers Alicia Markova and Anton Dolin (dancing to excerpts from Stravinsky's "Scènes de Ballet"), and singers Nan Wynn, Billie Worth, Bill Tabbert, and Dolores Gray.

Seven Lively Arts was the premier attraction to play Billy Rose's refurbished Ziegfeld Theatre, and Rose charged a $24 top price for the privilege of attending the opening night (but he did throw in a champagne reception). Lahr's material made it hard for even him to get laughs, though he did well enough with still another concert-baritone burlesque. This time the target was the operetta hero, which found Lahr decked out in a Lord Nelson uniform leading his roistering crew as they become inebriated singing a stirring drinking song called "Drink" ("Drink to Nelson Eddy before you faint/ And here's to J. J. Shubert, our patron saint").

Following the demise of *Seven Lively Arts,* Lahr again was faced with the problem of what to do next. Since his brand of comedy was unsuited to the book musicals of the time and revues were in a decline, he was forced to consider the possibility of doing a non-musical play. His opportunity came with the revival of *Burlesque.* Dating from 1927, when the leading roles were played by Hal Skelly and Barbara Stanwyck, this was a laugh-clown-laugh melodrama about the rise, fall, and rise again of a burlesque comic. Lahr tried it out in the summer of 1945 in Brighton Beach, New Jersey, then appeared in it the following summer (opposite Arlene Francis) in Greenwich, Connecticut. Sufficiently confident that he could play both the dramatic scenes as well as the comedy, he opened it on Broadway with Jean Parker later the same year. Obviously, there was a great deal of sentiment connected with Lahr appearing on Broadway in a play about burlesque, and he even directed the burlesque routines—including his old sketch about the "Dutch" cop and the gyrating dancer—that were added to the story. The result was a personal success for the comedian who, for the first time, was made aware of how effective he could be in reaching audiences not only through comedy but also through pathos.

Burlesque, which ran a year on Broadway, was a special kind of show that worked for Lahr at this particular period of his career, but there were no immediate opportunities to appear in other dramatic works. Early in 1949, the best he could get was to tour in the revue, *Make Mine Manhattan,* which had been a Broadway success featuring Sid Caesar. Lahr did most of the Caesar sketches and also added some of his own routines, including "Song of the Woodman" and the "Taxes! Taxes!" sketch, both from *The Show Is On.*

Two years later, Lahr returned to Broadway in another revue, *Two on the Aisle,* co-starring Dolores Gray. Though the show offered few new challenges, it did give Bert the chance to work with such talented people as Betty Comden and Adolph Green (responsible for most of the sketches and all the lyrics to Jule Styne's music) and Abe Burrows, here making his Broadway debut as a director. Two skits stood out. "Highlights from the World of Sports," written by Burrows, found Bert, as Lefty Hogan, a dim-witted baseball player being interviewed on television by an overly loquacious and fawning sports commentator (Elliott Reid).

COMMENTATOR: Now, Lefty, tell the folks, how long have you been donning those old spiked shoes?

LEFTY: Well, I've had these shoes about two, three years.

COMMENTATOR: I mean, how long have you been playing big-league baseball?

LEFTY: Well, let's see now . . . I started in 1920, and this is . . . 'Bout fourteen, fifteen years.

COMMENTATOR: If you started in 1920 that would make it more than twice as long as that.

LEFTY: Well, I ain't countin' the time the other side was battin'.

In Comden and Green's "Space Brigade," the year is 2492, and Bert is the intrepid Captain Universe, who leads his men in a variety of interplanetary adventures. In the scene, his rocket ship has just landed on an unexplored planet and the Captain appears in full outer space uniform, complete with cape and a helmet topped with an antenna. When one of his spacemen asks where they are, the Captain is quick to request a reading:

CAPTAIN: What's our stellar orientation?

FIRST SPACEMAN: 38 over 74 South by East Space.

CAPTAIN: Our orbitary retraction?

SECOND SPACEMAN: Zero over zero plus ten.

CAPTAIN: Our gravitational quotient?

THIRD SPACEMAN: 200 CC's centigrade.

CAPTAIN: Longi-space-itude?

FIRST SPACEMAN: 72 Space East.

CAPTAIN: Lati-space-itude?

SECOND SPACEMAN: 34 Space North.

CAPTAIN: Making our distance factor . . . *(He looks at his instruments.)* Men, as I calculate it—we're in Perth Amboy.

FIRST SPACEMAN: Captain Universe, haven't you made a slight error?

CAPTAIN: What's that? Are you saying that I have made an error? I, Captain Universe, leader of the Spa-a-a-ace Brigade, which asks no quarter and gives no quarter in its struggle against the malicious forces of interplanetary evil?

FIRST SPACEMAN: Oh, no sir, it's just that your dectolecticalter may be just off a bit.

CAPTAIN: Mmm . . . *(Shakes instrument.)* That's right. Shows you when you use cheap plutonium. This is two degrees off, which means . . . *(Does mental calculation.)* Men, this is the planet Venus.

With mission accomplished, the Captain instructs one of the spacemen to use the super-powerizer to telephone headquarters on Earth, 60 million miles away ("Make sure you reverse the charges"). Suddenly the brigade is surrounded by strange creatures with pointed green heads and pop-eyed goggles. They are, however, unseen by the Captain who calmly explains that his men should have no fear because Venus is uninhabited. When he does see the strangers he shrieks with fright, and when the spacemen's disintegrator gun fails to keep the creatures from advancing, the men make haste to follow their Captain's orders ("Run like hell!"). After the creatures unexpectedly run away, Captain Universe claims another victory for the Space Brigade, but they soon return with their green-haired, voluptuous Queen Chlorophyl. Instantly attracted to the Captain, the queen insists that he stay if she allows the other spacemen to leave. Agreeing to make the sacrifice, the Captain takes the willing queen behind a rock from which he soon emerges dazed, happy—and wearing pop-eyed goggles.

One song, "The Clown," gave Bert the chance to reveal something of the buffoon's secret yearnings. First he lamented the clown's lowly station ("Here am I condemned to be a japing, jeering jester,/ While others play Othello, King Richard, and Perle Mesta*"). Lahr then went on to show his versatility by imitating Rudolph Valentino ("Ah, my flower of the night, we shall ride across the pampas to my desert tent in Araby, where we will glide in gondolas through the jasmine-filled streets of Seville") and also a German-accented Queen Victoria ("Chentlemen, ve muss buy the Suez Canal as long as Chibralter shtands. Disraeli, Gladshtone, my minishters, you muss obey my vims! Remember, I am a kveen!"). Finally, though still frustrated in his attempts to convince others of his dramatic range, Lahr allowed, "But a clown must keep on clowning," and he ended the number emotionally crying, "Pagliacci-ooooooo."

If not exactly prophetic, "The Clown" did at least set the stage for Bert Lahr's next Broadway appearance. After making three undistinguished films, Lahr received what surely must have struck him as the most unlikely offer of his career, the role of Estragon in Samuel Beckett's existential play, *Waiting for Godot.* To producer Michael Myerberg, however, the offer was perfectly reasonable: he needed a "name," and he wanted a comic actor who could also play tragedy. Bert had come close in *Burlesque,* but *Waiting for Godot* presented the challenge of a lifetime. While admitting that he didn't understand everything in the play, Lahr was fascinated by his part, calling it the most satisfying thing he had ever done. Though reluctant at first to hazard so difficult a role, the actor had a great appreciation of Beckett's writing. He was also confident of the play's theatricality and the comic possibilities beneath the overlying theme of the survival of life as expressed through the relationship of two bums, Estragon and Vladimir (played in New York by E. G. Marshall).

Against his will, Lahr had to drop many of his familiar mannerisms and bits of business. He also had to limit his movements because the character he played was constantly complaining about sore feet. But without changing any lines, he did manage to inject certain bits that proved remarkably effective. It was, for example, Bert's idea to accompany the exclamation, "Ah!" with a raised index finger and a blank stare in order to indicate a total lack of comprehension. As in this exchange:

ESTRAGON: What do we do now?
VLADIMIR: I don't know.
ESTRAGON: Let's go.

*A reference to the famed Washington party-giver who was the inspiration for Ethel Merman's role in that season's hit, *Call Me Madam.*

VLADIMIR: We can't.
ESTRAGON: Why not?
VLADIMIR: We're waiting for Godot.
ESTRAGON: *(despairingly).* Ah!

When asked about the meaning of *Waiting for Godot,* Lahr once said, "Everybody has his own interpretation of *Godot.* At one point in the play, you think the tramps are waiting for God. But then Beckett would go off on another tangent. Then you knew it wasn't God. At the finish they are still waiting. Hopelessness. It was waiting for the best of life, and it never came. I think he meant the two characters to represent both sides of man. Estragon, my part, was the animal: Sex, Hunger, Eating, Sleeping. The other, Vladimir, was Suspicion, Inquiry, Intellect, always examining everything."

Under the direction of Herbert Berghof, *Waiting for Godot* was presented on Broadway in April 1956. As might have been expected, it was not a box-office success, running only 59 performances. But there was no doubt that it signaled an entirely new world of creativity for Bert Lahr. "It is Mr. Lahr's very personal intuition that counts," wrote Walter Kerr in the *Herald Tribune.* "He tugs in earnest desperation for a very long time at a shoe that is much too tight. Eventually he gets it off. When it comes off, rapture spreads over Mr. Lahr's face, over the great grey-blue background that suggests an empty universe, over the whole auditorium—and all in a single sigh. Delight is born before our eyes, flesh becomes something felt absolutely, and life—the kind of life that all the rest of us live—now begins."

To a new generation of theatregoers, the name Bert Lahr now became synonymous with that of *Waiting for Godot.* It firmly established him as a comic actor of extraordinary range and depth, and it opened up new creative challenges. On television he appeared in Shaw's *Androcles and the Lion* and in Molière's *School for Wives,* then returned to Broadway the following year in *Hotel Paradiso,* Peter Glenville's adaptation of Georges Feydeau's French farce. In the play, Bert was cast as Boniface, a hen-pecked, would-be seducer of his best friend's wife (played by Angela Lansbury in her first Broadway role), with most of the action taking place in the seedy hotel where he plans to have his rendezvous.

Throughout *Hotel Paradiso,* doors are flung open and banged shut, identities are mistaken, and stairs are constantly being used for one person to dash madly after another. The comic highlight occurs in the second act when Boniface, feeling a bit ill, leans against a hotel room wall. At that moment, a Peeping Tom hotel porter outside the room, curious to see what's happening inside, drills a hole through the wall stabbing Boniface and impaling him. Here Bert Lahr could make expert use of his wide assortment of hilarious facial expressions, registering not only pain but surprise, disbelief, terror, embarrassment, confusion, anguish, and even pleasure.

Unfortunately, farce—at least for modern audiences—seems to be appreciated in small doses or with songs and dances. Thus, despite favorable notices for both Lahr and the production, *Hotel Paradiso* could achieve no more than 109 performances.

In 1959, after having appeared in an existential comedy and a French farce, Lahr returned to the familiar precincts of a revue. It turned out to be his biggest failure. *The Girls Against the Boys,* in which Lahr was co-starred with Nancy Walker, also proved a shaky launching pad for the theatrical careers of Dick Van Dyke and Shelley Berman. A battle-of-the-sexes revue may have seemed a promising idea, but in execution it yielded only one memorable scene. Called "Hostility," the Arnold Horwitt–Aaron Ruben sketch of unrelenting domestic hatred was performed almost entirely in pantomime. According to John Lahr, in his book about his father, Bert enjoyed acting in the skit because it enabled him to get laughs exclusively through gestures and expressions without the aid of jokes, and because "silent family rage was something he knew from his childhood and his own adult temper tantrums."

The scene is a dreary-looking tenement room in which Bert, as a construction worker, and Nancy Walker, as his slattern of a wife, irritate, provoke, and enrage each other through a series of wordless, nerve-racking confrontations. Part of the sketch is devoted to the couple's dinner at the kitchen table. After opening the stove door to take out a stew—accompanied by clouds of black smoke—Nancy stirs it with a ladle and plops a portion onto Bert's plate. Bert sniffs the concoction as his wife opens two beer cans sending sprays of foam across the room. After one bite, Bert makes a face, picks up his plate, gets up from the table, opens the front door, and whistles for a neighbor's dog. But even the dog refuses to eat any of the slop. All the while, Nancy is quietly enjoying her dinner. When Bert indicates that there is no bread, his wife takes a loaf from the bread box and opens the table drawer on Bert's side to get the knife. Leaving the drawer open, she returns to her chair opposite Bert. After struggling to close the unwieldy drawer, her husband slams it so hard that it shoots out the other end of the table and hits Nancy in the stomach.

The scene ends when Nancy pulls down the Murphy bed with a loud crash. This awakens Bert, who has been napping on the sofa, and he stumbles across the room, drops his pants on the floor, flops wearily into bed, and falls asleep. Now in her slip, Nancy turns out the light and also gets into bed. There is a pause. "Eddie, Eddie," she calls, rousing her snoring mate. "You forgot something." Her husband grunts, leans over, and kisses her. Blackout.

In the fall of 1960, Bert Lahr joined the American Shakespeare Festival Company for a tour of *A Midsummer Night's Dream,* in which he played Bottom. Thus, Lahr became the only one among Broadway's classic clowns to play Shakespeare—and in the one Shakespearean role that Bobby Clark had always wanted to do. Lahr's ambition to play Falstaff, however, was never realized.

For *The Beauty Part,* which opened late in 1962, the talents of Bert Lahr were combined with the talents of our most literate of comedy writers, S. J. Perelman. The play was a somewhat buckshot attack on the foibles of those who profess to a deep concern for matters cultural but are only really interested in the fast buck, and was told through the adventures of a Candide-like innocent, played by Larry Hagman. *The Beauty Part* was actually a tour de force for Bert Lahr since it enabled him to show up as five different comic characters. As Milo Leotard Allardyce Du Plessis Weatherwax, the hero's father, he was a lecherous garbage disposal manufacturer (when surprised by his wife, he explained, "Now look here, Octavia, I just had to give our French maid a severe dressing down"). As Hyacinth Beddoes Laffon, he was the lady publisher of a lurid horror magazine called *Shroud.* As Harry Hubris, he was a double-dealing Hollywood producer whose offer of a $200,000 contract ends up as a $100 advance and a percentage of the Transylvanian gross. As Nelson Smedley, he was a cantankerous, red-baiting octogenarian. And as Herman Rinderbrust, he was a pompous judge reveling in the glare of television attention. Lahr thought it was the funniest material he had ever worked with, but the play lacked dramatic structure, and it was also badly hurt by a newspaper strike. The run was 85 performances.

Foxy, Bert Lahr's last Broadway show, had probably the most inaccessible tryout of any main-stem attraction. In July 1962, Lahr and his fellow actors journeyed to Dawson City in Canada's Yukon Territory, where the musical—which, fittingly, was set in the Yukon's gold-mining days—was presented as part of the first (and presumably the last) Gold Rush Festival. Financially aided by the Canadian government, the production was performed in the 501-seat Palace Grand Theatre and lasted seven weeks.

After undergoing extensive revisions, *Foxy* opened on Broadway in February 1964 under the banner of David Merrick. Though Lahr and his original co-star Larry Blyden were still in the show, the Robert Emmett Dolan–Johnny Mercer score was almost entirely rewritten, and there were new sets, costumes, choreography, and cast members. Loosely based on *Volpone,* Ben Jonson's classic tale of avarice, *Foxy* related how its titular hero, a gold prospector, outwits his double-crossing partners by making them think he will leave them huge sums of gold in his will. Lahr, then sixty-eight, made his entrance as Foxy staggering into a saloon wearing a beaver cap and a fringed buckskin coat, and with his left leg caught in a bear trap (an inconvenience he was totally unaware of, even though he had picked up the trap some eight miles out of town). Grandly tossing a fistful of lead pellets to his Eskimo guides, Foxy tells them, "Get yourself some chocolate-covered blubber." The prospector soon meets the wily Doc Mosca (Larry Blyden), who warns the saloon crowd that his new friend is a sick man given to fits. When Foxy obligingly offers a spectacular display of heaving, eye-rolling, and frothing at the mouth, Mosca is concerned that he might be overacting and mutters, "Enough." Foxy, however, has the stage and he will not be stopped. "Lemme alone," he tells Mosca, "I got a helluva finish." Later, playing a ruttish rogue, he tries to seduce a young bride (Julienne Marie), then becomes a mass of terror when she chases him. Also trying to escape from the bride's husband (John Davidson), Foxy dashes hysterically to the side of the stage. Then, with nowhere else to go, he simply shinnies up the proscenium arch, where he perches with a look combining fear with self-satisfaction.

Later in the show, Lahr masqueraded as a British lord, wearing a violet walking suit, deerstalker hat, and yellow leggings with black buttons. In this outfit he sang the self-descriptive "Bon Vivant," recalling such scenes of his youth as "Manchester and Dorchester and Chichester and Perth,/ Sailing on the Firth of Forth or is it Forth of Firth?" Though the musical itself garnered few admirers, all the first-night critics lauded Lahr's performance. "Bert Lahr should be preserved like a fine old wine, or in one," wrote Walter Kerr. "As the years go along, his tang gets headier, his lifted pinky gets daintier, his moose call to the great beyond gets mellower and mellower, and furthermore he is beginning to carbonate."

Bert Lahr gave his final Broadway performance on April 18, 1964, though two and a half years later he was back on the boards at the Ypsilanti Greek Theatre appearing in Aristophanes' *The Birds.* On December 4, 1967, while filming *The Night They Raided Minsky's,* he collapsed and died of a heart attack in New York. He was seventy-two.

The most insecure of all our great clowns, Bert Lahr ended his long stage career having dared more and succeeded in more areas of acting than any of his contemporaries. From Billy K. Wells to Beckett, Feydeau, Shakespeare, Perelman, and Aristophanes, he took on one challenge after another with the air of one who was always somewhat baffled by his own ability. As he commented at a Lambs Club dinner in 1964, "My career only goes to prove the truth of what Mark Twain once said, 'All one needs to gain success in life is to be ignorant and to have confidence.' I often wonder what would have happened to me if I had had a little confidence."

Beatrice Lillie

HER DARK HAIR is a closely cropped, mannish bob crowned by a black pillbox hat. Her eyes glint merrily with just the slightest suggestion of mischief. Her nose swoops down in a long arc, and her mouth is curved in a warm, inviting smile that follows the curve of her pointed chin. Here is a slim, gracious, elegantly gowned hostess making all of her friends feel right at home as she introduces the next entertainment on the program in a precise, British-accented voice. Then she hikes up her skirt and rollerskates offstage.

Throughout her long and acclaimed career, Beatrice Lillie remained preeminent in her ability to apply the slapstick with a regal hand. She was calm, cool and only occasionally ruffled, and the fact that she never wore herself out may explain, in part, why she never wore out her welcome. For Miss Lillie, above all other clowns, understood and practiced the comic art of detachment. Though she obviously enjoyed her own outlandish pranks, she never seemed to become emotionally involved in them; it was the distance between her and her material that lent enchantment to her performances as she set about puncturing all manner of social, artistic, and intellectual pretensions.

Beatrice Lillie never pandered to her audience. The roars of laughter that followed some hilarious bit of business would often provoke a stern look of disapproval which, of course, only convulsed her devoted fans even more. Theatregoers doted on her as they would a shrewd but rather barmy member of the social elite who had slid down from her perch to share with everyone her irreverent attitude about the vapid world she inhabited. Since her forte was satire, she could portray not only someone who is being put upon, such as a dowager struggling to maintain her dignity, but also someone who is doing the putting upon, such as a boorish socialite or an impertinent maid. No matter what she did, there was something irresistible in the way in which the upper crust was being scraped so deftly by one who, in Kenneth Tynan's words, "can convert a theatre into a throne room. Even at her most common she has the aristocratic touch."

That Beatrice Lillie was born into a lace-curtain Irish family in Toronto scarcely lessened that impression. As she wrote in her autobiography, *Every Other Inch a Lady*, "As far back as I can remember I always wanted to be slightly grand." In this goal she succeeded in fact as well as in fancy when her husband, Robert Peel, whom she married in 1920, inherited a baronetcy and she became Lady Peel.

Though her origin was just north of the United States, Beatrice Lillie first won success in London's West End, where she appeared for some ten years before making her Broadway bow in 1924. From then on, her career was divided almost equally between London and New York. Indeed, it was the very duality of her appeal that set Bea Lillie apart from all other stage comedians; for some forty years she was acclaimed with equal fervor on both sides of the Atlantic.

Born on May 29, 1894, Beatrice Gladys Lillie was the daughter of an emigrated Ulsterman who rose to be an official in the Toronto city hall. She first performed in public in Sunday school concerts, which so encouraged her ambitious mother that young Beatrice was enrolled in a private dramatic school. She made her professional singing debut as a member of the Lillie Trio, along with sister Muriel (who would later write songs for sister Bea) and her mother.

Because of Muriel's talent as both pianist and composer, Mrs. Lillie was determined that she have the best possible schooling. To her that meant only London, and it was there that Beatrice joined her mother and sister after they had been away almost a year. At the time, Bea's ambition was to be a "slightly grand" singer of heart-tearing ballads, which she felt would be the most rewarding career the theatre could offer. Auditioning for producer Andre Charlot's general manager, she therefore rendered such sentimental pieces as "Oh, for the Wings of a Dove" and Irving Berlin's "When I Lost You," but she found it impossible to hold the man's attention. As a sudden whim, she impulsively launched into the stirring strains of "God Save the King," which somehow made the general manager sit up if not stand up. As a result, he signed her to appear in Charlot's revue *Not Likely!*, currently running at the Alhambra Theatre. Miss Lillie's London debut took place in October 1914 when she joined Eddie Cantor (in his only London stage appearance) and the other performers some five months after the show's opening.

The Charlot revues were intimate affairs with small casts and a minimum of scenery, bearing such cryptic titles as *Samples*, *Some*, *This and That*, and *See-Saw*. Their legacy is that they provided early opportunities not only for Bea Lillie but also for Gertrude Lawrence, Jack Buchanan, Noël Coward, and Jessie Matthews. When Bea joined the cast of *Not Likely!*, she took with her the lachrymose ballads that she felt were proper for an elegant entertainer's repertory. Charlot, however, had other plans for her. Possibly because of Bea's short-cropped hairdo, he saw her as a "girl-hero" dressed in male attire expressing the sentiments of a proper gent yearning for a distant locale. That the locale invariably was situated in rural America, with the singer frequently in top hat and tails, was accepted then as part of theatrical tradition inherited from the music halls. Thus in Charlot's next offering at the Alhambra, *5064 Gerrard* (named after the backstage telephone number of the theatre), Bea swaggered out in a three-piece suit, surrounded by milkmaids, to reveal her homesickness in Irving Berlin's "I Want To Go Back to Michigan" ("That's why I wish again/ That I was in Michigan/ Down on the farm"). Her rendition was such a success that in spite of—or because of—the sight of this far-from manly impersonator longing to go back to a place for which she could hardly have had the slightest affinity, Bea Lillie's "girl-hero" image was established. In a succession of Charlot revues, she sang at least one number per show that would seem to have been more appropriate for Al Jolson in blackface than for Beatrice Lillie in jacket and trousers.

Because one revue followed another so quickly, Bea was appearing in *5064 Gerrard* while rehearsing the next Alhambra attraction, *Now's the Time* ("A Musical Timepiece in Two Hours and Ten Chimes"). The new revue's headliners were Clay Smith and Lee White, an American husband-and-wife song-and-dance team. In one scene, Bea, as the male partner, was paired with Miss White to offer Walter Donaldson's "We'll Have a Jubilee in My Old Kentucky Home," where the duo planned to settle down and enjoy "the corn and 'lasses served by Rastus."

There was little change in the Lillie image or repertory for the next two Charlot shows, *Samples* (which she joined about three months after the opening) and *Some* (in which she appeared for the first time with Gertrude Lawrence, who also served as her understudy). In *Some*, Miss Lillie formed a trio with Clay Smith and Lee White to sing the praises of the Old South in the Richard Whiting–Ray Egan ballad "(They Made It Twice as Nice as Paradise) And They Called It Dixieland." It was also during the run of *Some* that both Bea Lillie and Gertrude Lawrence (then billed as "Gertie") were fired (but quickly rehired) for cutting up backstage during one of Clay Smith's numbers. In tribute to the United States entry into the World War on April 6, 1917, Smith had written "America Answers the Call," which he introduced the very next night in front of the traveler curtain. His patriotic fervor struck Bea and Gertie as slightly ludicrous and they proceeded to march grimly back and forth behind the curtain, breaking up the rest of the cast and disrupting the performance.

Though she had long since abandoned her ambition to be a slightly grand

The sheet music cover of Beatrice Lillie's first Broadway show.

Beatrice Lillie (third from right) in "The Bus Rush" sketch in Noël Coward's 1928 revue, *This Year of Grace*. *

Walk a Little Faster, a 1932 revue, starred Bobby Clark, Paul McCullough, and Beatrice Lillie.

In *At Home Abroad* (1935), Beatrice Lillie tries vainly to order two dozen double damask dinner napkins from salesmen James MacColl, Reginald Gardiner, and Eddie Foy, Jr.

Beatrice Lillie in two scenes from her last musical, *High Spirits* (1964). In the first, she arrives on stage singing "The Bicycle Song"; in the second, she dances in her bunny rabbit slippers to "Talking to You."

singer of teary ballads, Bea Lillie was not content to remain a "girl-hero" either. She had discovered that she could make people laugh and she was now determined to become a clown—even though this talent occasionally got her into trouble. A 1917 Charlot revue called *Cheep* gave her her first real opportunity. In the first act she appeared in the customary male attire to render "When I Am with Her Again" (music by sister Muriel) as a Canadian soldier pining for his mother. She sang another song by Muriel, "Take Me Back to the Land of Promise" ("Back to the land of the ice and the snow"), which, since it was about Canada, had at least a tenuous plausibility—even while being interpreted in white tie and tails. She also wore the same outfit as she yearned to go back "Where the Black-Eyed Susans Grow," situated near a country farmhouse 'round a little country lane where a cornfield bride was waiting.

It was in the second act of *Cheep* that, in the words of Noël Coward, "her performance turned out to be an historic occasion because it was the first time the sleek and urbane Beatrice Lillie appeared in her true colors as a comic genius of the first order." The scene was a formal dinner concert given by a group called the Dedleigh Dull Quartette. As Coward recalled, "I forget the names of the other three performers but Miss Lillie, with her high-piled auburn hair and a green satin evening gown from the bosom of which protruded a long-stemmed chrysanthemum, will stay in my memory forever. She sang, in a piercing soprano, a straight popular ballad called 'Bird of Love Divine.' She sang this with apparently the utmost sincerity, but it did just occur to her during the second verse to prop her music up against the chrysanthemum. I believe I was still laughing hours later."

Others apparently had the same reaction. The number became the hit of the show, and the show itself had the longest run—483 performances—of any revue Bea Lillie appeared in for producer Andre Charlot.

Charlot's next attraction, *Tabs* (short for tableau curtain), followed *Cheep* into the Vaudeville Theatre. It was Bea's seventh Charlot revue in a row, and by this time she had become such a draw that her name was given billing just below the show's title. Her comic opportunities, however, seem to have been limited, with most of her appearances being in the familiar "girl-hero" vein. In "I Said Goodbye" (by Ivor Novello and Ronald Jeans), she was seen as a soldier who had a difficult time bidding farewell to his sweetheart. In another Novello-Jeans number, "I Hate To Give Trouble," she was the male half of a boy-flapper duet. And in Walter Donaldson's "I've Got the Sweetest Girl in Maryland," she again appeared as a formally garbed dandy to sing of returning to that state with the ring for his Mary's hand ("I know she'll meet me at the choo-choo/ With a minister man and a jasbo band"). Because of an accident while horseback riding, Miss Lillie was out of *Tabs* for two months, during which time she was replaced by her understudy, Gertrude Lawrence.

Bea Lillie's close association with impresario Charlot came to a temporary halt when she took a part in her first book musical, *Oh, Joy!*, which, in every respect other than the title, was the same show as the Jerome Kern–P. G. Wodehouse–Guy Bolton New York musical, *Oh, Boy!* (seems that the British sponsors felt that that title smacked too much of American slang for London audiences). Miss Lillie sang such early Kern charmers as "Till the Clouds Roll By" and "Nesting Time in Flatbush," and cavorted as a madcap actress in a confusing tale of marital misunderstanding among Long Island's socially elite. The production, which was billed as a "Musical Peace Piece," boasted in its program of having a male chorus made up of army veterans.

Bran Pie, which opened in August 1919 shortly after *Oh, Joy!* closed, returned Bea Lillie to the Charlot revue fold. The show's title gave the producer the chance to designate the scenes as "Dips," and its 400-plus performance run made it one of Charlot's longest-running revues. According to Miss Lillie, it was with *Bran Pie* that she decided to become more aggressive in achieving her goal as a comedienne. Curiously, Charlot still kept her pretty much restricted to singing take-me-back songs in male attire, but Bea's desire to do more comedy had become an all-consuming passion. One night, in sheer desperation, she simply wandered in and out of other people's sketches announcing, "Pardon me, you're wanted on the gramophone." Audiences may have found this amusing, but Charlot fined her for unprofessional behavior.

Among her songs, "That Wonderful Lamp" (by Nat Vincent and Jack Hulbert) gave her the opportunity to play a child wishing for Aladdin's lamp that would provide the answers to such questions as "Why father goes to Paris and why mother has tea with Mr. Harris." She also wore trousers to sing "Come Along, Mary" and "Take Your Girlie to the Movies," the latter a bit of practical romantic advice on how to avoid the girlie's interfering parents, "if you can't make love at home."

Though she had to leave the cast of *Bran Pie* during its run because of pregnancy, Bea returned to work as soon as she was able. Perhaps she should have waited a bit longer since her next appearance was in a tacky bedroom farce imported from New York called *Up in Mabel's Room*. The play barely lasted a month, but it did have the distinction of being the first of three non-musical plays that Miss Lillie acted in during her career. About this "Frivolous Farce in Feminine Foibles," the actress observed in her autobiography: "*Up in Mabel's Room* would be nothing I choose to be remembered for, so I refuse to remember it."

Bea happily returned to Charlot for an alphabetically all-inclusive revue, *A to Z* (anyway, it had 26 scenes), which was scheduled to open in October 1921. A bacterial infection, however, forced her to withdraw during rehearsals, and she was succeeded by Gertrude Lawrence.* Bea did eventually join the cast about a year after the show's opening and remained in it during the final month. Charlot, to help her get over her disappointment at having been unable to open in *A to Z,* added her to the cast of a revue then running in London called *Now and Then.* Her presence, however, did little to stimulate business and the show closed after only two months. Wasting little time, Charlot rushed Bea into *Pot Luck,* the next and more successful attraction at the Vaudeville, in which she appeared with Jack Hulbert and Herbert Mundin. Recalling her performance in this revue, critic J. C. Trewin wrote: "Her work is considered to the twitch of an eyelash, but she makes it appear spur-of-the-moment. Shoddy material is decorated until it gleams. A flick of the voice is a brushful of gold leaf. All the while she is whispering: 'Both of us know this is rather absurd. Just pretend for a moment it's rather good.' It is more; it can be marvellous, especially when she is preparing joyfully to puncture an irrelevent bit of drama. Behind the fooling is a calm, laughing intelligence; in Beatrice Lillie's presence the brash or the cocksure must wilt."

Bea Lillie's first revue without the guiding hand of Andre Charlot was *The Nine O'Clock Revue,* so-called because the curtain rose late enough to give fashionable Londoners enough time to dine and dance before going to the theatre. Bea's most memorable number was "The Girls of the Old Brigade" (music by sister Muriel), in which she led, with unaccustomed demureness, a bevy of ladies attired in only slightly exaggerated versions of what might have been seen in an 1880 Bank Holiday crowd at Hampstead Heath—all the way from skating outfits to bathing suits. (For no apparent historical reason, Miss Lillie's costume was an exact copy of Queen Alexandra's traveling dress.) Another Muriel Lillie contribution, "Susannah's Squeaking Shoes," which quickly became part of Bea's standard repertory, was a bouncing number celebrating the sounds made by the Sunday shoes of a lovable Tennessee mammy.

Since *The Nine O'Clock Revue* proved that Bea Lillie could win success without him, Charlot was concerned that she was lost to him forever. He had an idea that he was certain would not only appeal to Bea but also give him the chance of winning a new audience for his special brand of intimate, tasteful, witty revues: he decided to co-star the comedienne with two of his other top performers, Gertrude Lawrence and Jack Buchanan, in a revue that he planned to unveil in New York. Though there would be some original mate-

*Miss Lawrence wrote in her autobiography that it was in this show that she sang "Limehouse Blues" for the first time. The song's sheet music, however, bears the credit "Sung by Teddie Gerard," and an account in the *Daily Telegraph* also indicated that the song was added to the show for Miss Gerard.

rial, the bulk of the songs and sketches would be proven audience pleasers selected from the producer's London successes.

Andre Charlot's Revue of 1924 (the impresario was apparently taking no chances with a cutesy title) began its preparations with a good deal of apprehension about Broadway audiences appreciating British humor, particularly since *The Nine O'Clock Revue*, which had been imported the previous year without Bea Lillie, had been unable to attract a public despite favorable notices.* Nor was morale helped when, at a rehearsal, the show's American sponsor, Arch Selwyn, was heard to mutter, "It stinks!" But the opening at the Times Square Theatre on January 19, 1924, turned out to be a triumph for all, especially Beatrice Lillie and Gertrude Lawrence, who would from then on be Broadway favorites throughout their careers. Bea scored most notably with two numbers, Ivor Novello and Douglas Furber's "March with Me!" and Noël Coward's "There's Life in the Old Girl Yet." Both songs had already been introduced in London by a rotund comedienne named Maisie Gay, who sang the former in *A to Z* and the latter in Coward's first revue, *London Calling!*.

"March with Me!" ("March! March! April, May and June!"), something of a follow-up to "The Girls of the Old Brigade," found Bea as an imposing, if uncoordinated, Britannia, forever tripping over her feet and getting entangled with her shield and spear in a vain effort to keep anyone in the marching chorus line from taking over center stage. In "There's Life in the Old Girl Yet," Bea was seen as an aging but flirtatious soubrette singing of charms that are undimmed, even though "I'm as old as my tongue but much older than my teeth." As for Gertrude Lawrence, her strongest numbers were "Limehouse Blues" (inherited from Teddie Gerard) and Coward's "Parisian Pierrot." Following the unanimously enthusiastic critical reception, theatregoers began lining up at the box office in such numbers that the show remained in New York for 298 performances, then toured for twenty weeks.

The American success of *Andre Charlot's Revue of 1924* prompted Charlot to feature Lillie and Lawrence in a near-facsimile almost as soon as they returned to London in March 1925. Actually, this was the second edition of a show, then running at the Prince of Wales Theatre, which was also called, with confusing accuracy, *Charlot's Revue*. Lillie repeated "March with Me!" and "There's Life in the Old Girl Yet," and Lawrence again did "Limehouse Blues" and "Parisian Pierrot." Together they appeared as the Apple Sisters, Cora and Seedy, a ukulele-plunking, harmonizing duo, to offer a "Broadway Medley" in honor of their recent conquest. Their rendition was enlivened by Gertie imitating Sophie Tucker singing "Big Boy" and by Bea giving the Fanny

*Both "Susannah's Squeaking Shoes" and "The Girls of the Old Brigade" were sung in New York by Cicely Debenham.

Brice treatment to "I Wonder What's Become of Sally." They also did a sketch together called "Fallen Babies," a Ronald Jeans take-off on Noël Coward's play *Fallen Angels*. In the scene, Bea (sporting a curly blonde wig) and Gertie (with a lengthy curl descending from a baby's bonnet) played two infants in perambulators trading adult chitchat as they swigged gin that their nurses had secreted in their carriages.

By the end of the year it was back to New York for Lillie, Lawrence and Buchanan, who opened at the Selwyn in November as stars of *The Charlot Revue of 1926*. Though the reception was almost as warm as the first New York edition, the revue remained only about half as long as its predecessor. From *The Nine O'Clock Revue*, Bea revived "Susannah's Squeaking Shoes" and a sketch by Harold Simpson called "References," in which she played an impudent Cockney housemaid. Masquerading as her own mistress, who is no longer in need of her services, she is interviewed by a prospective employer about her maid's qualifications. "Has she unimpeachable credentials?" Miss Lillie looks shocked. "*That*," she says sternly, "I couldn't say. But I'm sure she has two of everything." From the West End *Charlot's Revue*, she and Gertrude Lawrence borrowed their "Fallen Babies" sketch, and from the first New York *Charlot's Revue* she brought back "March with Me!" ("By request," according to the program). Of the new material, the most successful routine was the burlesque of Anna Pavlova dancing "Les Sylphides," in which Bea invaded Fanny Brice territory as ballerina Wanda Allova dancing "Sealed Feet." Gertrude Lawrence joined Jack Buchanan for a song-and-dance treatment of an American song, "A Cup of Coffee, a Sandwich and You," and, as a solo, sang Noël Coward's "Poor Little Rich Girl" (originally done in the London revue *On with the Dance*).

After *The Charlot Revue of 1926* in New York, Miss Lillie and Miss Lawrence never again appeared together in the same production. For her first American musical comedy, Bea was signed by producer Charles B. Dillingham in the fall of 1926 to star in *Oh, Please!*, a limp vehicle about an actress named Lily Valli (Miss Lillie) who is somehow mistaken for the wife of a puritanical perfume manufacturer (played by Charles Winninger) and, quite naturally, causes all sorts of complications. The musical highlight of the Vincent Youmans–Anne Caldwell score was Bea's duet with Charles Purcell of "I Know That You Know," and for some barely comprehensible reason, a way was found to incorporate Bea's fashion pageant, "The Girls of the Old Brigade" (which critics compared unfavorably with "March with Me!").

There was nothing but praise, however, for Bea Lillie's performance. Wrote Brooks Atkinson in the *Times*, "Miss Lillie is a comedian with the divine spark, a mimic of the highest skill . . . She ridicules every traditional staple of musical comedy, everyone in the cast, everyone in the audience, and nearly everything she does herself . . . Miss Lillie remains ever so lightly detached. She is in it but not of it. The burlesque is so quick, so nearly imperceptive, that the audience has no sooner caught her meaning before she is three or four paces ahead to something quite as subtle and effervescent."

It was with *Oh, Please!* that Broadway critics first began to appraise Bea's facial characteristics. Her eyes, for example, were "mocking" (Frank Vreeland, *Telegram*), "the most serious in the world" (Gilbert Gabriel, *Sun*), and had a "menacing glint" (Alexander Woollcott, *World*). Her eyebrows were "quizzical" (Gabriel) and "eloquent" (Percy Hammond, *Herald Tribune*), she had a "remarkable Grecian nose with a tiny upward tilt at the end" (Vreeland), there was a "ribald upward quirk to her lips" (Vreeland), and her head had a "puckish tilt" (Woollcott).

About a year later, Bea Lillie returned to Broadway again under the Dillingham banner and again in a weak vehicle. As the comedienne explained, the story of *She's My Baby* "was one that predated *Uncle Tom's Cabin*, having to do with the rich old uncle arriving unexpectedly to meet the imaginary wife and baby that his improvident nephew had been telling him about, so *now* they must be produced somehow at twenty-four hours notice. Tilly, the maid, is talked into masquerading as the wife, the uncle is persuaded that the baby's been kidnapped, and—oh, you know how it turns out, and you can be sure who played Tilly."

Even though the score was by Rodgers and Hart, there was little about *She's My Baby* that pleased the reviewers—except Beatrice Lillie. To quote Gilbert Gabriel: "She yodeled and caroled and gulped and gesticulated in that cool, gingerly, determined way for which she is famous, the while she made her top notes splinter like glass around her. The imp in her always won out, leaped out of exaggerated finger tips and sly, over-haughty head tosses, until she was clowning for all her worth. Her game is limited, her tricks repetitious—but never mind that. She is the smartest female in Funnydom."

Bea Lillie was back on Broadway late in 1928 in *This Year of Grace*, a revue written by, directed by, and co-starring Noël Coward. Since the show was a new version of a successful production of the same name then running in London, songs and sketches written for other performers had to be refitted for the distinct personalities of Lillie and Coward. Miss Lillie put her own stamp on material—principally "Love, Life and Laughter," "The English Lido," and "The Bus Rush"—that had been associated with Maisie Gay, and Coward took over numbers introduced by Sonnie Hale (including "A Room with a View" and "Dance, Little Lady").

"Love, Life and Laughter" (which dated back to the London version of *Charlot's Revue*, not *This Year of Grace*) fround Bea as the "inviting, exciting" La Flamme, the toast of an 1890 French Bohemian café known as La

Chatte Vierge. There she is serenaded by an infatuated Noël Coward ("I want to smite you and beat you and bite you"), whom the coquettish siren entices by promising "Lovers may sip passion's wine from my lip." In "The English Lido," set in 1902, Bea appeared as Daisy Kipshaw, a self-important, publicity-hungry Channel swimmer in a garish red bathing suit who strikes ludicrous poses for newspaper photographers while insisting on her love of privacy, addresses the admiring bathers as if they were her subjects, and signs autographs with a condescending flourish. When one timid young man refuses to part with his autograph book, Daisy tears it out of his hand, scribbles her name in it, and flings the book triumphantly over his head.

Perhaps Bea's most memorable sketch in *This Year of Grace* was "The Bus Rush." Done entirely in pantomime—except for one word at the end—the scene is set at a bus stop on a London street corner. We first see Miss Lillie entering from stage left as a grim-faced Lady in a frumpy dress topped by a turban-like hat with feathers, clutching three large parcels and three balloons presumably intended for the kiddies back home. Mustering her middle-class dignity as if it were a shield, she takes her place beneath the bus sign, only to have other people somehow manage to edge ahead of her in the queue. The Lady, of course, is clearly not one to tolerate such ill-mannered behavior, and she imperiously brushes past anyone who may have been so foolhardy as to try to take advantage of her. Soon a motor horn is heard offstage right which the people in the line follow with their eyes as the "bus" moves to stage left and everyone dashes in that direction to get on board. But the bus is apparently full, the bell rings, and all return to their former positions. This business is repeated, with the Lady becoming increasingly harried though no less determined to preserve her own fragile self-respect. When, after another unsuccessful attempt to board a bus, a motor horn is again heard offstage right, everyone gets the bright idea of dashing to the right. But again the bell rings on the left and the people reverse their direction, though with no better luck than before. By this time, the Lady is completely bedraggled, her parcels crushed, her balloons deflated, her hat knocked to the side of her head, her knees wobbly. She may be ready to collapse but she is not defeated. With head held high, she manages to summon enough strength to lift one finger and call, with all the haughtiness she can muster, "Taxi! Taxi!"*

In addition to material that had been previously seen in London, *This Year of Grace* also had songs and scenes that Coward wrote especially for Miss Lillie. "World Weary" showed her as an office boy munching an apple while sitting on a high stool and yearning to "get right back to nature and relax." "I

*W. C. Fields's sketch about subway travel in the *Ziegfeld Follies of 1921* (see page 77) was an earlier commentary on the frustrations of public transportation.

Can't Think" found her recalling an embarrassing occasion when she tripped and fell at the feet of a handsome stranger ("I feel as he had not the strength to raise me/ He might at least have joined me on the ground"). (Those with sharp memories were aware that this was a burlesque of "I Don't Know," Gertrude Lawrence's recollection of a more successful flirtation which she had sung in *Andre Charlot's Revue of 1924.)*

In "Lilac Time," Lillie and Coward appeared in a two-character spoof of treacly Viennese operetta (a form of musical entertainment that Coward himself would soon become involved with in *Bitter Sweet).* The curtains part to reveal Miss Lillie seated on a bench in a moonlit bower primly reading a book. Enter Mr. Coward, swooning with passion. Unseen by his inamorata, he steals behind the bower, rolling his eyes fervently, then sings of his ardent yearnings—only to have Miss Lillie react with a convulsive jump of fear. Having unwittingly ruined the romantic mood, she is overcome with mortification and struggles to repair the damage by assuming a suitably worshipful gaze that will both recapture the rapture and atone for her behavior. She also shoots a withering look at the audience for daring to laugh at her predicament. Her song cue comes to her rescue, leading the lovers into such observations as "Birds are chirrupping love's sweet song," "Blossoms are o'er the lea," and "That's why its lilac time under the chestnut tree." The scene ends with the radiant couple about to exchange a discreet kiss, which is thwarted when the bench on which Miss Lillie is sitting topples over and lands her on the ground.

Critical praise was bestowed on *This Year of Grace,* with reviewers again paying special attention to Bea Lillie's facial and body language. In the *Journal,* John Anderson noted the effect Bea had on her audience "simply by crooking her finger, dropping a wayward eyebrow, or lifting an uncertain voice." According to Arthur Pollock in the *Brooklyn Daily Eagle,* "No other person can do so much that is amusing with a wave of the hand or the raising of the eyelid." Robert Littell commented in the *Post,* "The scenes depend on Miss Lillie's ability to make us roar by moving one finger or one eyelash." And the *Sun*'s Richard Lockridge summed up his views by observing, "The charm of her method lies in its delicacy, as in her personality one of the most tangible qualities is restraint. She never broadens her effects and never points. It is more devastating to see one corner of her mouth twitch than it is to see some comedians fall downstairs."

In 1930, after making her initial appearance in New York's vaudeville mecca, the Palace, Bea Lillie returned to London for her first West End engagement in over five years. Again under the aegis of Andre Charlot, she was starred in *Charlot's Masquerade,* a revue that was the premier attraction at London's newest playhouse, the Cambridge. Most of the material was original, with sketches by Ronald Jeans and lyrics by Rowland Leigh set to the music of some nine different composers. Miss Lillie did, however, take along Rodgers and Hart's "A Baby's Best Friend" ("Is her mother"), which she had introduced in *She's My Baby*, and she also set the scene for a sketch called "Hollywood" that had been performed in the recently opened third edition of the *Garrick Gaieties*. It was done in two tableaux: the first as people imagine Hollywood to be—a den of debauchery and violence—and the second as it really is—a den of debauchery and violence. In addition, Bea imitated the celebrated monologuist, Ruth Draper ("In this little sketch, ladies and gentlemen, I want you to imagine *far* too much"), sang a song while roller skating around the stage, did a clog dance wearing galoshes, skipped gaily around a Maypole as Queen of the May, led a crew of gallant sailors in a nautical number called "Lady Clara," and took part in a scene showing the influence of Negro music on a group of formally clad Mayfair socialites. Though the reviewers welcomed Bea warmly and the show itself garnered favorable notices, *Charlot's Masquerade* was somehow never able to find an audience, and it closed after fewer than one hundred performances.

In 1929 the intimate and sophisticated *Little Show*, starring Clifton Webb, Libby Holman, and Fred Allen, was the nearest thing Broadway had seen to the two Charlot revues and *This Year of Grace*. *The Second Little Show*, without stellar performers, fared poorly, and the producers were convinced that the only one who could give *The Third Little Show* the necessary quality and appeal was Bea Lillie. Opening in June 1931, with comic actor Ernest Truex co-starred, "The Aristocrat of All Revues" (as it was billed) was deemed superior to the second edition but still not quite up to the first. And it certainly did not help matters when the far more enthusiastically received revue, *The Band Wagon*, opened just two nights later.

The Third Little Show provided the stage debut for one of Bea Lillie's most closely identified specialties, "There Are Fairies at the Bottom of Our Garden." Her friend Ethel Barrymore had suggested that this authentic nineteenth-century parlor song, composed by concert singer Liza Lehmann, would be perfect for the Lillie treatment. Though initially Bea was unconvinced, she did try it out in her night-club act, and it became such an instant hit that she was determined to find room for it in her very next revue.

In the scene, Miss Lillie appeared as a formally gowned concert singer with a long pearl necklace twined around her neck and a huge ostrich fan that she waved at the audience. She may have presented an excessively regal appearance, but the singer was still unable to resist injecting her own mocking asides while delivering the fey, childish sentiments of the ballad. Following her description of the impish behavior of the fairies, she then sang:

The King is very proud and very handsome,
And the *Queen*—well, can you guess who that may be?

After advising her listeners, "Shhh, now, this'll kill you," she then sang the concluding lines:

She's a *lit*-tle girl all day
But at night she steals away—
Well, it's meeeeeeeeeee! Yes, it's me!

With an exultant "Wheee!," she whirled her string of pearls so that it looped out from her neck, then kept on whirling about her until it fell at her feet.

In other sketches, Bea appeared in "The Late-Comer" as a movie patron who flounces about telling everyone within earshot how the picture turns out; showed up as an English spinster in an Apache café in Paris who natters away with her friends as a murder is committed at the next table; and, in a holdover from *Charlot's Masquerade*, again gave her impression of Ruth Draper. Noël Coward provided Bea with two choice items. The first was the song, "Mad Dogs and Englishmen" ("Go out in the midday sun"), a putdown of British inability to adjust to tropical climate. Bea sang it wearing a solar topee and tropical whites while seated in a rickshaw surrounded by black natives. The second Coward contribution was a sketch, "Cat's Cradle," which he had written as a scene for two women but which was changed for the revue to a woman and a man. The setting is the back view of two suburban houses where two neighbors, Miss Tassell (Bea Lillie) and Mr. Mawdsley (Ernest Truex), come out of their respective houses to chat over their dividing backyard wall. Each owns a cat, and it is Miss Tassell's wish that her Walter "should get intimate with some other cat of his own station." When the lady suggests Mr. Mawdsley's Minnie, Mr. Mawdsley is shocked:

MISS TASSELL: Well, you know life's life all the world over, and you can't escape from it. You're standing in the way of Minnie's happiness.

MR. MAWDSLEY: Happiness! Oh, Miss Tassell! How can you? If I thought Minnie harbored such ideas after her life here with me, I'd never forgive myself.

MISS TASSELL: Facts are facts, you know. I've seen your Minnie walking up and down with ever such a wistful look in her eye.

MR. MAWDSLEY: I don't know what to say. I don't really. I feel quite strange.

MISS TASSELL: Of course, in the circles I move in, the facts of life are discussed quaite quaite openly. False modesty is so—er—middle class, don't you think?.

MR. MAWDSLEY: False modesty is one thing, Miss Tassell, loose thinking is another.

MISS TASSELL: I beg your pardon.

MR. MAWDSLEY: Granted as soon as asked. Minnie is not as other cats. Her life has been very secluded. I merely don't care to picture her in any—er—peculiar situation.

MISS TASSELL: Well, I only hope she won't lose her fur as she gets older. Repression is a very bad thing. If in your narrow-mindedness you refuse to allow her to fulfil her natural destiny. . . .

MR. MAWDSLEY: And what if I don't consider your Walter to be Minnie's natural destiny?

MISS TASSELL: That, Mr. Mawdsley, is what the French would call "un autre pair de souliers"!

By 1932 Beatrice Lillie had appeared in seventeen stage productions, only three of which were book musicals and one was a non-musical farce. Nevertheless, the directors of the prestigious Theatre Guild felt that she was the perfect actress to do full justice to the role of Sweetie, the nurse, in George Bernard Shaw's play, *Too True To Be Good.* The play, which contained Shaw's opinions on various subjects including vegetables, war, inoculation, and the British army, had one special problem as far as Bea Lillie was concerned. As she wrote in her autobiography: "I hadn't the faintest notion what my lines meant." The reviewers, however, welcomed her as if she knew all too well. But her struggle to fathom the unfathomable and the restrictions of saying every line as written with no chance for ad-libs, left Bea with one desire during the play's run: she wanted to appear again in a good revue.

While her next Broadway show, *Walk a Little Faster*, was undeniably a revue it was also undeniably not a very good one. According to Bea Lillie, the chief pleasure it gave her was the chance to appear on the same stage with Bobby Clark, one of her idols. The subtle technique of a Lillie and the raffish approach of a Clark did not, however, mesh too well, chiefly because of the general weakness of their material. Among the sketches Bea appeared in were S. J. Perelman's "Scamp of the Campus," a college romp set in 1906 in which she played coquettish Penelope Goldfarb (wearing a flat straw hat and a shirtwaist with ballooning sleeves); "Frisco Fanny," in which she was the proper proprietor of a Yukon saloon and Bobby was a lecherous prospector; and Robert MacGunigle's "The Girl Friend," in which she was Nancy Fixit, a catty friend visiting a stage star in her dressing room after an opening-night performance. She also demolished torch singers with her excessively mawkish treatment of the pop song, "I Apologize." But *Walk a Little Faster*, which opened late in 1932, did earn its place in the history books for two reasons: it was the first show in which Bea Lillie appeared as a roller-skating grande dame, and it was the show in which Vernon Duke and E. Y. Harburg's "April in Paris" was introduced.

Beatrice Lillie returned to London a year later to be in *Please* (not to be confused with *Oh, Please!*), her sixteenth and final revue under the auspices of Andre Charlot. Her starring partner was the diminutive, acrobatic comic, Lupino Lane, and most of the material consisted of warmed-over American songs and sketches. From *Walk a Little Faster* there was the "Frisco Fanny" scene, "The Girl Friend" scene, and the Lillie version of "I Apologize." From *The Band Wagon* there was "Hoops," which Fred and Adele Astaire had introduced and which was now being done by Lillie and Lane, and from *Flying Colors* there was "Louisiana Hayride," which became the finale in *Please*. The show's run at the Savoy was only about three months.

As far as output was concerned, the Broadway theatre never had two more prolific producers than the Shubert brothers, who jointly sponsored more than one hundred musicals, both book shows and revues. In 1933 they set about presenting a series of tasteful, melodic, and witty revues, all at the Winter Garden and all featuring some of the most gifted performers on Broadway. The first three in the series were the *Ziegfeld Follies of 1934*, with Fanny Brice and Willie and Eugene Howard, *Life Begins at 8:40*, with Bert Lahr and Ray Bolger, and *At Home Abroad*, with Beatrice Lillie, Ethel Waters, and Eleanor Powell.

At Home Abroad, a self-described "Musical Holiday," was an elaborate production that used the "theme" of an around-the-world trip through which it could offer all manner of comedy scenes, dance routines, and songs both risible and romantic (by Arthur Schwartz and Howard Dietz). Staged and designed by Vincente Minnelli, the revue opened September 19, 1935, to a generally enthusiastic press.

Within the global framework of the show, Miss Lillie could be seen in a variety of climes, nationalities, and situations. In London she was Mrs. Blogden-Blagg trying vainly to order two dozen double damask dinner napkins at a department store but only succeeding in getting herself—and the salesmen—totally confused as she tried to relay her tongue-twisting request.* In Paris, as a Moulin Rouge poster come to life, she sang the excessive praises of "Paree" while perched halfway up the backdrop wearing a huge, elaborately plumed black hat, a gold satin evening gown, black net stockings, and arm-length gloves as she dangled a cigarette in a foot-long holder. In Monte Carlo a sketch by

*Though the routine became identified with Miss Lillie and she repeated it in the Bing Crosby film, *Doctor Rhythm*, it was written by Dion Titheradge for British comedienne Cicely Courtneidge. Miss Courtneidge first performed it in 1928 in the second edition of a London revue, *Clowns in Clover*.

Howard Dietz found her as Sonia Polonariskaya, the première ballerina of the Imperial Ballet, who moodily waits in her dressing room for her lover to show up. As she waits, she sadly reminisces with her maid, Babushka (Vera Allen), about her unhappy life in Russia.

SONIA: From the time I was two and a half years old when I first entered the Imperial Ballet, I had to walk four miles through the snow to get to rehearsals because I could not afford a droshki. I walked all the way on my toes to make my toes tough. I have tough toes, haven't I, Babushka?

BABUSHKA: You have the toughest toes in the whole ballet.

SONIA: You were with me when I danced the mazurka for the Czarevitch Alexis and when I did "The Dying Drake" for the Little Father himself. Then came the revolution, remember? But I could not dance for the mujiks. I could not face the mujik. So what did they do? What did they do? One night they dragged me out of bed. Out of bed, mind you. And I had to hurry home, gather my few possessions, and flee. I trekked from St. Petersburg to Moscow. From Moscow to Omsk. From Omsk to Pinsk.

BABUSHKA: I know.

SONIA: Just a second. Finally I trekked all the way across the border where we artists met and again formed the Imperial Ballet. But I am tired now, Babushka, and I cannot trek any more. Because you can't teach an old dog new treks.

When, just before her lover finally arrives, Sonia despondently refuses to go on stage that night, her manager (Reginald Gardiner) pleads with her not to disappoint the audience. Sonia is adamant:

"I will never dance again."
"You won't dance?"
"Don't ask me."

In the Swiss Alps (the first-act finale), Bea joined Herb Williams in a yodeling song, "O Leo," then led the entire cast—each member tied to another—as they seemingly scaled the towering Matterhorn during a snowstorm. In Vienna she entered down a spiral staircase into a chandeliered ballroom as Madcap Mitzi singing to her smartly uniformed admirers, "I'm the toast of Vienna and most of Vienna can boast they've been host of the toast of Vienna." In England again, she appeared at a railroad station to bid an emotional goodbye to her lover after they have enjoyed a romantic weekend in the country. Vowing never to see each other again lest it ruin the perfection of their love, they tearfully part, and the man goes off to catch his train. Somehow, however, he misses it and when he spoils the mood by returning, Bea bops him over the head with his suitcase.

The Orient was represented by a scene in a Japanese garden where the ladies of the ensemble enter with cherry blossoms in their lacquered black wigs, wearing low-cut gowns suggesting kimonos, and twirling parasols. With bowed heads and expressionless faces, they take their places in a kneeling position from which, in unison, they musically urge male tourists to "Get Yourself a Geisha, a gay little geisha," as the most accommodating guide to the wonders of their country. After the ladies have described the pleasures they offer and reach the line, "And sleep in the forest of the cherry blossoms," suddenly from their midst comes the recognizable voice of the heretofore unrecognized Bea Lillie with the bit of advice, "It's better with your shoes off." The audience's surprise at suddenly discovering the star of the show produced one of the biggest howls of the evening.

At Home Abroad provided as festive a way possible for theatregoers to enjoy being abroad at home, with Bea Lillie again the darling of the Broadway critics. As John Mason Brown put it in the *Post:* "She is her cool, collected, immaculate self. Hers is the happiest of zany gifts inasmuch as she knows when to let well enough alone. She never pushes her fooling too far or seems to coarsen it, no matter how broad or how unblushing is the thing that she may have been called upon to sing or to say. Her technical means are so tidy and apparently so 're-fined' that they become doubly funny when they are contrasted with the rough-and-tumble quality of her material." But for 1935 *At Home Abroad* was something of an anachronism in offering a fun-filled romp through foreign lands during a period of international upheaval. Even after dropping its top price from $4.40 to $3.30 (when it moved from the Winter Garden to the Majestic Theatre), it had a run of fewer than 200 performances.

Beatrice Lillie was again starred by the Shuberts in *The Show Is On,* their fifth and final Winter Garden revue of the mid-1930s. This time the "theme" was concerned with all manner of musical and non-musical entertainment, and the show provided Bea and her co-star, Bert Lahr, with some of the best material of their careers. Vincente Minnelli was once more on hand to direct and design the production, and an impressive array of Broadway talent was tapped to write the songs and sketches.

Bea's mocking humor was notable in a wide variety of numbers. She was an excessively animated rhythm singer, GoGo Benuti, complete with a rhythmically bobbing spray of orchids, tearing through a Rodgers and Hart piece called, appropriately, "Rhythm" ("There's rhythm in the treetops/ There's rhythm under the sea/ There's rhythm in this heart of mine"). Part of the rendition called for her to string together snippets of pop tunes in which the last word of one song became the first word of another. Bea even tossed in "The Star-Spangled Banner" with the stern command, "Up everyone." (She also did this number in the movie *Doctor Rhythm.*)

Another fondly remembered routine in the show found Bea doing a turn as a vaudeville soubrette perched precariously on a crescent moon singing the mockingly gooey sentiments of Herman Hupfeld's "Buy Yourself a Balloon" ("Take a trip to the moon/ And stay up there and croon"). As she sang, a crane swung the moon out over the heads of theatregoers in the first few rows, from which position Bea coquettishly bestowed garters on favored baldheaded gentlemen seated below her. She also kidded Josephine Baker's appearance in the *Ziegfeld Follies of 1936* by performing an exotic dance that had her slithering all over a white divan until she awkwardly fell to the floor; became a stripper in a burlesque of burlesque in which she tossed stockings, a brassiere, and false hair from behind a row of six fan dancers; and, in "The Reading of the Play," was seen as a visiting French actress in the 1890s who, when told that the drama she is to appear in is about a man and a woman, has the single comment: "Too much plot."

In "Mr. Gielgud Passes By," a sketch by Moss Hart, Bea had one of the show's comic highlights. Possibly inspired by "The Late-Comer," in which Miss Lillie had played a talkative woman in a movie theatre in *The Third Little Show,* Hart's skit was more immediately prompted by the coincidence of having both John Gielgud and Leslie Howard starring in *Hamlet* in New York at the same time. There are two scenes: the first in the dressing room where Gielgud (played by Reginald Gardiner) voices apprehension about appearing before a rude opening-night Broadway audience, and the second showing the first six rows of the orchestra as well as part of the stage. The play is in progress as Mrs. Slemp (La Lillie) parades down the aisle and is prevented by an usher from taking her seat until the end of the first scene.

MRS. SLEMP: Don't make me livid, young man. My friends will tell you that when I'm livid I'm not pretty—so don't make me livid. Will you unhand me or do you want me to call Herbert Bayard Swope? He'll be livid, I warn you.

USHER: I'm only an usher.

MRS. SLEMP: What did you think I thought you were, the Ballet Russe? *(The usher leaves.)* That's better. There's time enough for that sort of thing when the revolution comes and not before.

Not knowing quite where her seat is, Mrs. Slemp squeezes past the people sitting in one row and proceeds back through the one behind it. Paying no attention to the play, she turns around, waves to a friend, then sees another friend ("How divine to see you! I love your dress—almost as much as the ones they're wearing this year"). After some more searching, Mrs. Slemp finally locates her seat but continues to talk to friends about where they will meet after the performance. When a man behind her tries shushing her, she turns on him with the one withering word: "Communist!"

After listening a bit to the play, Mrs. Slemp calls to a friend, "Don't you just hate *Hamlet?* It's too revolting. They all talk like those dreadful New Dealers." When Gielgud appears on stage to recite the lines, "O, that this too too solid flesh would melt," it only reminds Mrs. Slemp of a "divine reducing belt," and his line about "customary suits of solemn black" only serves to recall a black suit she'd bought at Bergdorf Goodman. With that, Mrs. Slemp flings her fur cape over her shoulders and it somehow lands on the head of the man at her left. Glaring at him as if the man were a thief, she brusquely retrieves the cape and tucks it under her shoulder strap.

GIELGUD: To be or not to be . . .
MRS. SLEMP: Oh, I know this!

Mrs. Slemp proceeds to let everyone know that she knows by reciting aloud the first few lines, but only succeeds in getting the words mixed up. She prattles on about what she had once done in a school play, then puts her finger to her lips to silence a friend who hasn't said a word. As she does so, she bumps the arm of the man at her right and knocks his program to the floor. While the man searches for his program, she holds him down so that she can talk to the woman on the other side of him. Gielgud, now well aware of Mrs. Slemp, begins talking louder and louder in order to be heard over her voice. When she tries speaking to two friends at once, the actor is practically screaming his lines, and Mrs. Slemp motions him to be quiet.

Thoroughly exasperated, Gielgud leaps off the stage and stands in the aisle fuming at Mrs. Slemp. She jumps when she sees him.

GIELGUD: Allow me to introduce myself. I am John Gielgud and I intended, in my curious English fashion, to play Hamlet in this theatre tonight.

MRS. SLEMP: Not here! Really! Actors in the audience. One must draw the line somewhere. I could spit!

Since it is obvious that Mrs. Slemp has absolutely no intention of stopping her mindless chatter, Gielgud gets a bright idea: he offers her two tickets to Leslie Howard's *Hamlet*—only to be told that it was Howard who had given her the ticket to *his* production.

The success of *The Show Is On* sparked the production of *Happy Returns*, Bea's next London revue. Producer Charles B. Cochran became convinced that if the cool mockery of Beatrice Lillie would work so well with the boisterous clowning of Bert Lahr, it would also work with the equally boisterous clowning of Bud Flanagan and Chesney Allen, a popular music-hall act. For her first West End revue in about five years, Bea repeated some of the best-received songs and routines that she had done in *The Show Is On*—"Rhythm," "Mr. Gielgud Passes By" (retitled "The Rival Hamlets"), "The Reading of the Play," and "Buy Yourself a Balloon." In addition, she brought over "Paree," "The Toast of Vienna," and "Get Yourself a Geisha" from *At Home Abroad*. Though, according to Cochran, Bea stopped the show almost every time she was on stage, Londoners apparently found her less than ideally matched with her raffish co-stars, and *Happy Returns*, unhappily, lasted only about a month.

Set to Music, in 1939, reunited Beatrice Lillie with Noël Coward for their first collaboration since *This Year of Grace*, though this time Coward did not appear in the show. Opening in New York early in the year, it was another example of a revue put together with material that had largely been performed in a previous show, in this case Coward's 1932 West End revue, *Words and Music*. Miss Lillie inherited three of the best items on the program—"Mad About the Boy" (from Norah Howard), "Midnight Matinee" (from Ivy St. Helier), and "Three White Feathers" (from Doris Hare).

"Mad About the Boy" was a four-part song in which a Society Woman, a Housemaid, a Girl of the Town (that's what she was called in the New York program), and a School Girl each sings of her infatuation with the same screen idol. Bea played the School Girl who finds it hard to concentrate on her homework while drooling over a photograph of the star of such Hollywood epics as *Strong Man's Pain* and *Can Love Destroy?*

The first-act finale, "Midnight Matinee," found Bea as Mrs. Rowntree, a tireless but frazzled organizer of a charity pageant being presented before royalty, in which various society matrons contribute their impressions of "Bygone Enchantresses." As each enchantress is introduced, descends three steps from a platform in the rear and performs an appropriate turn, all manner of crises occur—props fail to work, sequences are reversed, lines are blown, cues are mixed up—which require Mrs. Rowntree to take direct action. When, after performing a sinuous dance, Salome mistakenly leaves the salver with the head of John the Baptist on the stage, Mrs. Rowntree, while still in the wings, tries hooking it off with a walking stick. After that fails, she bravely strides on stage, picks up the platter, and strides off. When Joan of Arc is almost choked by her long blue cape that somehow gets caught on one of the columns on the platform, Mrs. Rowntree crawls onto the platform on her hands and knees, unhitches the cape, and crawls back to the wings. When Lady Godiva's name is announced and there is neither Lady nor horse, a scuffling and neighing sound is heard offstage, followed by the shocked voice of Mrs. Rowntree: "There! Look what it's done now!" For the scene's grand tableau finale, Lady Blessington trips on the top step, falls headlong, and knocks Marie Antoinette's wig off. Once more Mrs. Rowntree bravely comes onstage to be of help, but she is pushed aside by the other participants as they frantically try to hide the mishap from the audience. To contribute to the confusion, the scene ends as two cherubs fly in on wires from opposite sides of the proscenium and collide in mid-air.

"Three White Feathers," the third Lillie number held over from *Words and Music,* found Bea as a former provincial actress seated in an automobile with her husband, a Guardsman, on her way to being presented at court. The young woman is so nervous that she makes rude remarks to to spectators lining the procession outside Buckingham Palace:

HE: You really must be a little more dignified.

SHE: Dignified? With three white feathers perched on me 'ead . . . head. I feel like one of the horses in *Cinderella.*

HE: Keep calm.

SHE: It's all very fine for you. You're used to it. Here am I a dancing soubrette for fifteen years suddenly shoved on in the Palace scene without a rehearsal.

Thinking back on her humble origins, she remembers her father's pawnshop with affection, then sings of her uneasy climb in scaling the social ladder ("Today it may be three white feathers, but yesterday it was three brass balls").

As for the new material in the show, Miss Lillie got things off to an hilarious start by making a grand entrance as Brünnhilde (it was her idea) seated side-saddle on a white horse in which her Valkyrian high notes were interspersed with commands to her unsteady steed. From then on the audience was hers. In "Opening Night," Bea appeared as a stage star in her dressing room thoroughly miserable with her lot, even though she is surrounded by attractive young men proffering gifts. As she seizes the presents avidly, she morosely comments as, one by one, the suitors offer her rubies ("Such *cru*-el stones, aren't they?"), diamonds ("Lovely, but so cold"), and pearls that had belonged to one young man's mother ("And did they make her happy?"). Once the men have left, the disillusioned actress bares her soul in "I'm So Weary of It All" (a similar sentiment to Coward's "World Weary," which Bea had sung in *This Year of Grace*), in which she sheds tears over her glamorous but empty life and yearns to return to the simple pleasures of her childhood.

Set to Music also marked the first revue in which Bea sang "I Went to a

Marvelous Party," a song inspired by a party Noël and Bea had attended on the Riviera at which they had been expected to entertain. Bea sang it in the show wearing slacks, a fisherman's shirt with padded puffed shoulders, about ten strands of pearls, a large sun hat, and dark glasses. After describing the antics of the guests—Laura getting blind on Dubonnet and gin, Cyril ripping off his trousers and jumping in the sea, seventy-four-year-old Elsie swinging upside down from a glass chandelier, etc.—she ended the refrain with the near-hysterical stamp of approval, "I couldn't have liked it more!"

Curiously, despite his strong contributions, Noël Coward was generally taken to task by the reviewers for not providing Miss Lillie with fresher material. But the comedienne, as usual, could do no wrong. "No one whose pursuit is laughter is her equal in poise," wrote John Mason Brown. "No one whose target is absurdity can put dignity to such a devastating use . . . She stalks the absurd like a setter in the field. When she scents it, she does not content herself with pointing. She snatches the gun out of the hunter's hand and fires it with that deadly aim which is uniquely hers. Never has her marksmanship been more side-splitting or her invention brighter than it is in *Set to Music.*"

During the World War II years, Beatrice Lillie spent most of her time entertaining British troops at bases throughout the world. She did, however, find time for two London revues, *All Clear* in 1939 and *Big Top* in 1942. *All Clear,* in which she co-starred with Bobby Howes and Fred Emney, took full advantage of such recent Lillie successes as "I'm so Weary of It All" and "I Went to a Marvelous Party." She also dipped back a bit further for "O Leo" (from *At Home Abroad*) and Noël Coward's "Cat's Cradle" sketch that she had performed with Ernest Truex in *The Third Little Show.* This time, although Miss Lillie again played opposite a man, Bobby Howes appeared in drag as Miss Mawdsley. The revue took note of the war primarily by beginning in a blackout and ending in a cheerful blaze of light.

In *Big Top* Miss Lillie was co-starred with Cyril Ritchard and Fred Emney. Her most memorable number was a satirical lament, "Wind 'Round My Heart" (by Harry Parr Davies, Barbara Gordon, and Basil Thomas), in which, as a mournful figure in gray, she gloomily expressed the agony she still felt about a brief but torrid romance. Part of the show's problem was that it was basically a small-scale intimate revue that was forced to accommodate the large-scale dimensions of His Majesty's Theatre, with the result that much of the wit and satire of Herbert Farjeon's sketches was lost on the audience. Like *Happy Returns* in 1938, *Big Top* was presented by Charles Cochran.

World War II may have been raging in 1944, but Broadway producer Billy Rose did not let that stop him from reopening the Ziegfeld Theatre with an unsparingly lavish revue called *Seven Lively Arts.* Though at first Rose wanted to star his former wife, Fanny Brice, in the show, her refusal paved the way for Bea Lillie's return to Broadway after an absence of five and a half years. For the second time her co-star was Bert Lahr, but Bea was clearly the major attraction; so enthusiastic was her welcome by the first nighters that the proceedings had to be delayed over five minutes (this was partly an expression of sympathy since Bea's only son, who was in the Royal Navy, had been reported missing in action).

Rose made sure that *Seven Lively Arts* was extravagant in every way. Not content with having just two major stars, he also signed Benny Goodman, Alicia Markova, and Anton Dolin. The songs were by Cole Porter, Bea's sketches were by Moss Hart, the staging was by Hassard Short, and even Igor Stravinsky was represented by excerpts from his new work "Scènes de Ballet." The first-nighters showed up in formal attire and paid $24 a ticket (this also entitled them to attend a champagne reception), but despite a generally approving press the show did not remain longer than 183 performances.

Bea Lillie introduced two mild comic specialties, "When I Was a Little Cuckoo" (a song of reincarnation) and "Dancin' to a Jungle Drum" (subtitled "Let's End the Beguine"). She did a bit better with the sketches. They included a travesty on the play *Angel Street*, with Bea as the long-suffering Mrs. Manningham whose reaction to her husband's malevolent behavior is to gaze at him adoringly and coo, "You're so good to me"; a putdown of the addlebrained British aristocracy, "There'll Always Be an England," with Bea as Lady Agatha, the hostess of a wartime canteen, gamely trying out her own version of American slang; and "Ticket for the Ballet," a variation on Hart's own "Mr. Gielgud Passes By," in which Bea played a garrulous ticket buyer who holds up the line behind her by trying to recall the name of the ballet she wants to see (she thinks it's called "Shurok"), and even attempts her own version of "The Afternoon of a Faun" to help her remember.

Better Late, an intimate revue credited mostly to Leslie Julian Jones, brought Beatrice Lillie back to London in 1946. The opening number started things off brightly. The curtain rose to reveal a row of smartly gowned ladies waltzing to the strains of a languid melody. The ladies wear large black masks with elaborate filigree designs and as they dance they deftly remove the masks with graceful, sweeping gestures. One lady, in her turn, makes a properly graceful, sweeping gesture, but the mask refuses to come off. As the others withdraw into the wings, the lady remains still frantically tugging and pulling at her mask. Finally, a vigorous wrench does the trick and—surprise!—it's Beatrice Lillie!

The revue combined both old and new material. Bea, swathed in mink, did an opera take-off, singing the "Waltz Song" from Edward German's *Tom Jones*, and once more offered a sendup of emotional vocalists with her rendition of sister Muriel's "To Hold You in My Arms." She also joined two other singers in baby clothes to squeal the misery of being "Triplets," an Arthur

Schwartz–Howard Dietz number previously sung in *Between the Devil,* and she again offered her toast to "Paree," the Schwartz-Dietz tribute she had introduced in *At Home Abroad.*

In assessing Bea Lillie's performance, Harold Hobson wrote in the *London Sunday Times:* "When in a song she comes upon a note she cannot reach she shrugs her shoulders in a way that suggests the entire world of music is beneath her regard. No one else in the theatre can so charmingly, so accurately, and so triumphantly as Miss Lillie give the impression that the grand, the beautiful, and the exquisite are unworthy of her notice. This sort of thing has to be done with the most delicate of precision. The least touch of exaggeration falsifies it. There is no falsification at all in Miss Lillie's art, which, in *Better Late,* is seen at its entrancing best."

By common consent, *Inside U.S.A.*, in 1948, gave Bea Lillie her most enjoyable Broadway revue since *The Show Is On* almost twelve years before. Her co-star was the bland, amiable comic Jack Haley and her material was furnished by Arthur Schwartz and Howard Dietz (songs) and Arnold Auerbach and Moss Hart (sketches). Schwartz also doubled as producer.

Just as Schwartz and Dietz's *At Home Abroad* had allowed Miss Lillie to scamper merrily through various foreign ports of call, so *Inside U.S.A.* (which claimed to have been "suggested" by John Gunther's famous book) now gave her a chance to take comic advantage of a variety of locales confined to a single country. The tour begins in Pittsburgh, where a formally clad choral society is seen onstage and is presently joined by Miss Lillie as the primly efficient music director carrying her music stand which she places at center stage. After stiffly bowing to the audience, she turns to face the singing group, raises her arms, stamps her right foot twice, then lowers her arms, examines the music, and blows a pitch pipe. Reassured, she again raises her arms, stamps her right foot twice, and the singers lift their voices in the musical invitation to "Come, O Come" to the industrialized, polluted metropolis of "Peetsburgh" ("Come and commune there, That's *if* you can find your way").

Arnold Auerbach's sketch "A Song To Forget," is somehow fitted into the show's format by having Bea, as a movie fan in Chillicothe, Ohio, dream that she is the inspiration of three classical composers in a film biography. As the notorious femme fatale, Mme. Lapis de Lazuli, she enters her study as students serenade her from the street below. "Oh, they *are* tiresome, these young studenten," she sighs. "So many baritones. Last night at the Katzenhoffenheuerathskeller they were at my table until six this morgen, drinking beer from my slipper." (Chopin enters the room.) "So nu, Freddie, wie geht's? Written anything lately?" Though Chopin is enflamed with passion, Mme. de Lazuli resists his advances because she is convinced that to yield would ruin his talent. "What is a woman?" she asks rhetorically. "A mere glockenspiel in the orchestra of life." Now suitably inspired, Chopin composes the "Polonaise" for her. After Chopin leaves, Liszt enters the study. I'm all dried up, he cries to Mme. de Lazuli, who simply lowers her shoulder strap, thereby providing the proper inspiration for the composer to create the "Hungarian Rhapsody" on the spot. He leaves and Tschaikowsky enters, but he is no more successful romantically than either Chopin or Liszt. When the two other composers return, the three men get into such a heated discussion about their music that they totally ignore the now fuming object of their desire.

"Massachusetts Mermaid" found Miss Lillie as the mischievous creature perched on a rock describing herself as a "piscatorial Episcopalian," and singing of her rather awkward love life. For the stop at New Orleans (which ended the first act), Bea is the Mardi Gras queen who has her hands full—in the song, "At the Mardi Gras"—avoiding the advances of over-eager celebrants ("A pirate got irate/ He seemed to desire it/ By any means").

Perhaps her most celebrated sketch was Moss Hart's "Better Luck Next Time," located geographically "Just Off Broadway," in which Bea played Gladys, the outrageous maid of a nervous stage star. The scene is the dressing room before the opening-night performance, and Gladys not only whistles and sings, much to the actress's annoyance, but she is constantly making unnerving comments. When the stage manager pops his head into the room and wishes the star good luck, Gladys is appalled. "Whenever the stage manager wishes you good luck at eight-fifteen on opening night," she tells the actress, "you can figure we close in three weeks. Never saw it fail, dear. Wouldn't give you two cents for this clambake now." She further nettled her employer by telling her she has worked out her name in her numerology book, and it came out "Titanic." ("According to my book, you should have been dead about three years ago. A little more rouge on your cheeks, dear. You do look a bit pale.") After consulting her ouija board and communicating with John Wilkes Booth for more dire predictions, the board flies out of her hands. Gladys looks shocked. "That doctor in Boston was all wrong," she tells the actress. "You *are* pregnant!" Blackout.*

Inside U.S.A. ran for nine months at the New Century Theatre, then toured for over a year.

In July 1952, Bea Lillie began what was supposed to be a three-week summer-theatre engagement in an intimate revue, *An Evening with Beatrice Lillie.* Made up mostly of the comedienne's choicest songs and sketches and performed by a cast of five (including Reginald Gardiner), the entertainment proved so successful that Bea toured in it for nine weeks. This led to a Broad-

*Curiously, a variation of this sketch had shown up two years earlier in Miss Lillie's London revue, *Better Late.* It was then called "Out of This World," with Dennis Waldock credited for adapting the Moss Hart script.

way run at the Booth Theatre beginning in October. Though scheduled for five weeks, it ran for eight months, toured again for ten months, ran in London at the Globe for five months (with young Leslie Bricusse in the cast), and toured Britain until the fall of 1955.

Familiar material in the first act included the sketches "The Girl Friend" (retitled "A Star's First Night"), first performed in *Walk a Little Faster*, and "References," first performed in *The Nine O'Clock Revue*, plus the song "Wind 'Round My Heart" from *Big Top*. The second act was devoted to Bea's musical specialties, including three that had been done previously only in night clubs and on radio. These were the Muriel Lillie–Nicholas Phipps "Maud" ("We're all of us just rotten to the core"), in which Bea, bolstered by brandy, coolly tells off a friend ("Maud, you're full of maggots and you know it"); a song about another Maud, "Come into the Garden, Maud," in which a booming baritone serenades Miss Lillie whose only reaction is a look of startled disbelief; and the quayside lament of a jilted traveling companion who wails, "They've slapped a sticky label on my heart, 'Not Wanted on the Voyage.' "

The Broadway critics were full of superlatives. Both the *Times*'s Brooks Atkinson and the *Journal-American*'s John McClain dubbed Bea "the funniest woman in the world"; the *World-Telegram & Sun*'s William Hawkins came up with "one of the greatest clowns of all time"; the *Daily News*'s John Chapman, though appalled at the $6.00 top price for so modest a show, wrote that she was "the equal of Chaplin as a pantomimist and, at the same time, the most disarming of people"; and the *Post*'s Richard Watts, Jr., apologetically commented, "It isn't paying her enough of a compliment to say that she is the foremost lady clown."

Bea Lillie's opportunity to appear in a *Ziegfeld Follies* unfortunately came at a time when that form of extravagant entertainment had become sadly out of date. Though Ziegfeld had died in 1932, rights to the title had been granted the Shubert brothers for three editions (1934, 1936, and 1943) and to other sponsors in 1956. Tallulah Bankhead starred in that production which closed out of town before reaching New York. The following year, rights went to two new producers who somehow managed to convince Bea to appear in it. The epicene comic, Billy DeWolfe, headed the large supporting cast.

Bea was responsible for the idea of her best skit, a pantomime scene called "Milady Dines Alone." Into an elegant French restaurant on a snowy night comes an imperious lady with snowflakes clinging to her broad-brimmed, feathered hat which almost covers her eyes. Still wearing her elbow-length white gloves, she endeavors to read the menu by holding it at arm's length while peering at it through the feathers in her hat. The meal itself turns out to be a struggle for dignity preservation since it consists of stalks of asparagus that droop everytime she tries to bite into them, an elusive cob of slippery corn that keeps rolling off the table, and a lobster whose claws the lady carefully manicures with a pair of garden shears. Except for occasionally dipping her gloved hands into a finger bowl, nothing distracts her from her haughty, diligent consumption of everything on the table.

Unfortunately, the rest of the material proved as limp as the asparagus, with such reminders of past glories as a sketch in which she played an obnoxious airline stewardess unnerving passengers by singing "I've Got a Feeling I'm Falling," and a song which allowed her to again ride a crescent moon over the audience while distributing garters.

According to Bea Lillie, novelist Patrick Dennis had her in mind as well as his own aunt when he wrote *Auntie Mame.* Miss Lillie did eventually get to play the title role in the Jerome Lawrence–Robert E. Lee stage version when, in June 1958, she took over the part in New York from Greer Garson (who had succeeded Rosalind Russell) and then opened in the play in London in September. Her return to the West End after an absence of three and a half years helped turn the play into one of the major attractions of the season.

Another role that, according to Miss Lillie, was written with her in mind was that of Mme. Arcati, the medium, in Noël Coward's 1941 play, *Blithe Spirit.* Her wartime commitment to entertaining the troops, however, made it impossible for her to appear in the original company. Bea did play the part twenty-three years later, but it was in the musical version, called *High Spirits,* written by Hugh Martin and Timothy Gray. Coward made sure to put his own stamp on the production by serving as director (though Gower Champion was called in during the tryout) and by rewriting some of the scenes. Thanks to a receptive press and public, the musical remained in New York for almost a year.

The relationship between Beatrice Lillie and Noël Coward, though they were friends throughout most of their adult lives, was usually stormy when they worked together as star and director. With regard to *High Spirits,* Bea was vexed that her part wasn't large enough and that Noël didn't fully appreciate her contributions to the show. For his part, Noël was convinced that Bea couldn't take directions or even remember her lines. Once the show opened, however, they resumed their normally close friendship.

The story of *High Spirits* is concerned with a happily married man (Edward Woodward) whose deceased first wife (Tammy Grimes) is accidentally materialized by Mme. Arcati during a séance, and the comically awkward situations that result. Bea managed to find all manner of business to help enliven her part. She made her entrance riding a bicycle with a rearview mirror, then proceeded to take the bicycle clips from around her legs and snap them around her arms like bracelets. She wore an incredible array of garish outfits, including a mandarin robe, harem shirts and trousers, space boots, and endless ropes of pearls and other jewelry. In her funniest scene, she was shown relaxing in her bedroom wearing a frilly nightcap, an old-fashioned nightgown with a bright yellow flower pinned to it, and bunny slippers with long ears that flopped as she moved. Gazing at her adored ouija board, she sang, in "Talking to You," all about what a happy medium she was to have such a wonderful companion. Feeling courageous, she even attempted a softshoe dance full of mincing, modest steps. Suddenly emboldened to try something more daring, she turned into a madly leaping and pirouetting ballerina, with arms waving wildly above her head. As the audience roared its approval, she took a mocking bow, full of exaggerated gestures of appreciation and farewell.

It was in *High Spirits* that Beatrice Lillie made her final stage appearance on February 27, 1965. She was then nearing seventy-one. Her years since then have not been comfortable. The woman whom Noël Coward once called "a gallant, shining and indestructible star" would have reason to laugh her sardonic laugh at that word "indestructible." At this writing, Miss Lillie has reached ninety, but for at least the past ten years has been suffering both physical and mental breakdown. After giving up her home in New York in 1974, she was placed in a nursing home near Henley-on-Thames, England, where she still lives.

Victor Moore

As THEIR CAREERS have shown, great Broadway clowns did not necessarily have to be lovable or sympathetic. Even those whose humor stemmed from their inability to cope with life's adversities were usually careful to avoid being too pitiful or downtrodden lest emphasis on these qualities had the effect of inducing tears rather than laughs.

The major exception was Victor Moore. Everybody loved Victor Moore. He was Broadway's most endearing, wistful, pathetic, and gullible clown. His stage personality, which was something of an extension of his offstage personality, was that of the unassuming, humble, bumbling, trusting little man who is incapable of dealing with any of the constant misfortunes that beset him. What helped keep the characterization comic and not tragic was, first of all, Moore's appearance. Moore looked as funny as he did woeful. The features of his pudgy face were set in an oversized head which, particularly in his later years, seemed to fit uncomfortably atop a dumpling-shaped body supported by short, skinny legs. His walk was more of a wamble and his voice was a cracked and plaintive quaver. Though his normal expression was that of a man either perplexed or worried, he made people laugh simply because he never allowed his discomfort to be taken too seriously. Nor did he ever allow his performances to become too broad; always a model of control, they were never cluttered by excessive grimaces, gestures, or business.

In order to extract the greatest possible humor from the Moore persona, librettists and playwrights were forever dreaming up occupations for him that were totally at odds with his character. Since there is nothing basically funny about a meek incompetent, Moore was usually cast in roles that found him in occupations customarily associated with more aggressive, self-assured types. This was even true of his first major role, in the 1906 musical *Forty-five Minutes from Broadway,* in which he played a former prizefighter and racetrack tout. Later, he depicted characters engaged in such unlikely pursuits as bank-robbing, rum-running, and burglary. Between 1931 and 1941, Moore's most successful Broadway years, he was seen as a Vice President of the United States (in both *Of Thee I Sing* and its sequel, *Let 'Em Eat Cake*), a gangster *(Anything Goes)*, a United States ambassador *(Leave It to Me!)* and a United States Senator *(Louisiana Purchase)*. Even productions for which he was sought but in which he did not appear, such as *Red, Hot and Blue!* and *Mexican Hayride*, would have found Moore playing, respectively, a convict and a fugitive con man.

Not only was Victor Moore's comedy enhanced by its contrast with incongruous occupations, it also benefited from being contrasted with the personalities of other actors. This was particularly true of the seven book musicals in which he was teamed with William Gaxton. Gaxton's brash manner and nervous energy provided a perfect balance to Moore's phlegmatic style, and the two enjoyed a series of stage triumphs that made them unquestionably the outstanding musical-comedy team of their day.

One aspect of Moore's career that set him apart from the other comedians covered in these pages is that almost all of his Broadway appearances were in book musicals rather than in revues. Revues, of course, have always provided clowns with the chance to act in a variety of roles, to expand their techniques, and to experiment with and develop new characterizations or routines. Moore, however, seldom did any experimenting. Once he had learned his lines and established the way he would do a particular role, he rarely changed anything. His aim was always to give the best performance possible by opening night, then repeat it without variation for the length of the show's run. Moore never found this boring; on the contrary, this approach was to him the height of professionalism in the theatre.

Victor Frederick Moore was born in Hammonton, New Jersey, on February 24, 1876. From his earliest years he wanted to be an actor, an ambition that was encouraged by his parents. When he was a young boy, his family moved to Boston, where his father opened a restaurant. By the time Victor was fifteen he had joined an amateur stock company, and two years later he made his first professional appearance in Boston as an extra in a local production of *Babes in the Woods*. Following that inauspicious debut, he joined a number of traveling stock companies where he gained valuable experience in a variety of roles, including villains.

The season of 1896–97 marked Victor Moore's first two Broadway appearances—though neither part was large enough to win any notice. His initial role, that of a waiter in a sentimental romance called *Rosemary* (subtitled "That's for Remembrance"), was so small that he wasn't even listed in the program until the play went on tour. The production, which ran 136 performances at the Empire Theatre, had a remarkably distinguished cast that included John

Drew, Maude Adams, Ethel Barrymore, and Arthur Byron. Later in the season Moore appeared briefly as a servant in Sardou's *Spiritisme*, starring Maurice Barrymore. By the end of the century, however, Moore was anxious to branch out as a vaudeville comedian, and he placed an ad in the *New York Clipper* announcing his availability: "A Neat, Refined and Novel Act Suitable for Any Audience. Boy impersonations and Parodies that Are Catchy and Up-to-Date. Everything New and Original. My Own Author."

After an unimpressive beginning as a "monologist, mimic and descriptive singer," Moore bought a vaudeville sketch that soon made him a headliner in the world of two-a-day. Though the actor had first appeared in it in 1902 with a singer named Julia Blanc, after his marriage the following year to Emma Littlefield, he toured in it exclusively with his wife. The sketch set the style and tone of the Moore stage personality throughout his career. "Change Your Act, or Back to the Woods" consisted mostly of banter between a vaudeville couple and a burly stage manager who tells them they're through unless they come up with some new routines. Victor then improvises a new monologue about a newsboy and his dog, Emma tries out an original song-and-dance number with the aid of the pit orchestra, and the manager is so impressed that he welcomes the team back as headliners. In the scene, Moore could tug at audiences' heartstrings at the cruel way he has been handed his notice, and his eventual success gave everyone the supreme pleasure of seeing the humiliated little man come out on top. Moore, however, never felt that need to follow the sketch's advice. "Change Your Act" was the only sketch he and his wife ever appeared in in vaudeville.

In the summer of 1905, Sam H. Harris, George M. Cohan's business partner, saw the act at Hammerstein's Theatre and became so enthusiastic about Moore that he made Cohan see the sketch, too. At that time Cohan was starring in *Little Johnny Jones* and had just finished writing a musical comedy, *Forty-five Minutes from Broadway*, that was tailored for the talents of singer-comedienne Fay Templeton. Victor Moore was the actor he wanted to play her leading man. Miss Templeton, then forty and nearing the end of her Broadway career, presented a rather dumpy figure on the stage. Moore may have been ten years her junior, but since the curly-haired, cherubic-faced performer, who stood 5′7″, was already on the chubby side, it helped make the romantic attraction believable. Most important, perhaps, is that Moore had already shown in "Change Your Act" the kind of personality the role required. For though Kid Burns, the part that he would play, had been a prizefighter and a racetrack gambler and tossed off wisecracks with the slangy accent of the New York streets, he was essentially the same kind of vulnerable, decent, sentimental person that Moore could portray so well.

No one, however, could have predicted that he would score such a re-

Victor Moore and Fay Templeton in the dramatic scene from George M. Cohan's 1906 musical, *Forty-five Minutes from Broadway*, in which Miss Templeton tears up the will that would have made her rich. *

In a scene from the 1926 Gershwin musical, *Oh, Kay!*, Gertrude Lawrence's popping champagne-bottle cork frightens Frank Gardiner, Oscar Shaw, Victor Moore, and Sascha Beaumont.

A tense moment in *Heads Up!* (1928) involving Jack Whiting, Robert Gleckler, and Victor Moore. *

In the Pulitzer Prize winning *Of Thee I Sing* (1931), Vice Presidential candidate Alexander Throttlebottom (Victor Moore) meets Presidential candidate John P. Wintergreen (William Gaxton) and assorted politicians.*

The stars of the 1934 Cole Porter hit, *Anything Goes:* William Gaxton, Ethel Merman, and Victor Moore.*

The sheet music cover of Victor Moore's 1938 success.

William Gaxton and Vera Zorina try to frame Victor Moore, as an investigating United States Senator, in *Louisiana Purchase* (1940).

sounding hit, especially in a production written for someone else. Yet when *Forty-five Minutes from Broadway* opened January 1, 1906, at the New Amsterdam Theatre, Moore had all but eclipsed the show's star in the eyes of both critics and public. According to the uncredited *Times* reviewer, "He has most of the good lines in the play, but it may be said to his credit that many others take on a quality of humor simply because he has a natural and unctuous manner of speaking them. And his occasional lapses into sentiment have a genuine ring."

The musical—whose title refers to the proximity of New Rochelle—concerns the problems that arise when a skinflint millionaire dies without leaving a will and his property is assigned to his nephew (played by Donald Brian). When the nephew (plus fiancée and prospective mother-in-law) and his secretary, Kid Burns, arrive in New Rochelle, the Kid falls in love with Mary Jane Jenkins (Miss Templeton), the dead man's housekeeper. Though Mary eventually discovers a long-lost will leaving everything to her, the Kid tells her he cannot marry for money and Mary dramatically tears up the will. There were only five songs in the score, but three of them—"So Long, Mary," "Mary's a Grand Old Name," and "Forty-five Minutes from Broadway," which Moore introduced—turned out to be among the most enduring pieces in the Cohan canon.

Quick to capitalize on the popularity of *Forty-five Minutes from Broadway,* Cohan next wrote *The Talk of New York* as something of a sequel in which he again hoped to feature Fay Templeton and Victor Moore. Miss Templeton, however, demurred, and in his second major Broadway appearance Moore received solo star billing. In the new variation, the actor again played Kid Burns, though this was the only holdover character from *Forty-five Minutes from Broadway.* Here Burns is a slangy racetrack gambler with a talent for picking longshots. He makes friends with a wealthy socialite named Wilcox, and, much to the displeasure of the man's wife and son, falls in love with the Wilcox daughter, Geraldine. Eventually he wins over the family when he exposes the son's girl friend as a gold digger. The dialogue once again gave Moore a full quota of wisecracks. Sample: When urged to bet on a particular horse because "Sunbeam ran second twice," the Kid remains unimpressed. "So did William Jennings Bryan," he scoffs. The score contained two especially catchy numbers, "When a Fellow's on the Level with a Girl That's on the Square" (sung by Moore) and "When We Are M-A-Double R-I-E-D" ("H-A-double P-Y we'll be"). The final scene in *The Talk of New York*—as if to reinforce its tenuous link with *Forty-five Minutes from Broadway*—takes place in New Rochelle where the Kid and his bride have gone to settle down. The show proved another popular Cohan attraction and ran for 157 performances.

A second and far less successful attempt to cash in on the appeal of Victor

Moore in the Kid Burns character was *The Happiest Night of His Life*—though Cohan had nothing to do with it and the character Moore played was named Dick Brennan. But there was no question that it was essentially the same role—a street-smart racetrack gambler with a sentimental streak—who this time is concerned with joining a friend in taking the friend's uncle for a night on the town.

The show's three-week run early in 1911 was Victor Moore's last New York appearance for fourteen and a half years. It did not, however, end the Kid Burns connection. Part of the time Moore toured in two plays whose leading male characters were again obvious ripoffs of the tough-talking, tender-hearted character. *Shorty McCabe,* written by Owen Davis, even had ads that used the shameless slogan, "Shorty McCabe Has Kid Burns Beaten from the First Curtain," and *Patsy on the Wing* sought to suggest *Forty-five Minutes from Broadway* in its tale of a poor girl who inherits a large estate. Moore also toured in a Guy Bolton–P. G. Wodehouse musical called *See You Later.* And he and his wife were always available to play "Change Your Act" on the vaudeville circuits which they did until 1924.

By the time the comedian re-emerged on Broadway in the fall of 1925, it was as if he were beginning a new career. He was heavier, his personality more self-effacing, his voice even more of a tearful bleat. He waddled more noticeably, and his comically ungainly shape was emphasized by trousers that came only to the tops of his ankle-high boots. His once curly locks were gone and what hair he had was gray (though he would soon have it dyed). Moore's first return appearance, in which he co-starred with Otto Kruger, was in *Easy Come, Easy Go,* another play by Owen Davis. Moore was cast as an experienced bank robber who introduces young Kruger to the trade. After a robbery, the two hide out at a health farm, where they have a number of comic adventures and where true love appears to make Kruger return to the path of honesty. The play proved a hit and ran for six months. In appraising Moore's performance, the unsigned *Times* critic wrote, "He has a quality both wistful and pathetic, and he has a choking voice at moments of emotion that should make those laugh who have never laughed before." It was a comment that could have been made at any time in Moore's later career.

Victor Moore's ascendancy in musical comedy, however, may be traced largely to the long-forgotten Bolton–Wodehouse musical in which he had toured. In 1926, seven years after *See You Later,* the libretto-writing partners were vainly searching for the right comedian to support Gertrude Lawrence in her first American musical, *Oh, Kay!,* which would have a score by George and Ira Gershwin. Suddenly the writers recalled that, despite the failure of *See You Later* to open on Broadway, it did have a respectable tour chiefly because of the performance of Victor Moore as a philandering husband from Elmira,

New York, who comes to New York City for an extramarital fling. The image of the hesitant, sad-faced, roly-poly comedian in a totally unlikely role provided a comic twist that, they reasoned, could work as effectively in *Oh, Kay!* Here, however, the contrast was in casting Moore as a bootlegger. According to Bolton and Wodehouse, neither of the show's producers, Alex A. Aarons nor Vinton Freedley, had ever seen Moore on stage before, but they agreed to sign him solely on the writers' strong recommendation.

Once rehearsals had begun, Aarons and Freedley were appalled. Moore's tentative, low-key manner of reading his lines was far from the style of comic aggressiveness that the producers had been accustomed to in Broadway funnymen. They became so concerned, in fact, that just before the Philadelphia tryout they virtually promised another comedian, Johnny Dooley, that he would replace Moore. Moore's performance as the bungling rum-runner, however, turned out to be one of the highlights of the show. From then on, the comedian was an Aarons and Freedley favorite, appearing in three of their subsequent musicals *(Funny Face, Hold Everything!*, and *Heads Up!)*, as well as two that Freedley produced on his own *(Anything Goes* and *Leave It to Me!)*.

Oh, Kay! was set in a summer house in the fictitious locale of Beachampton, Long Island, where, unknown to its absent owner, Jimmy Winter, the cellar is being used by bootleggers to store whisky. The harmless and hapless trio was made up of a British duke (Gerald Oliver Smith), his sister, Kay (Gertrude Lawrence), and Shorty McGee (Moore). To escape suspicion after Jimmy has returned to make preparations for his forthcoming marriage, Kay and Shorty pass themselves off as his newly hired maid and butler.

In one of the show's comic scenes, Kay and Shorty are called on to serve lunch on the terrace for Jimmy (played by Oscar Shaw), his fiancée (Sascha Beaumont), and her father (Frank Gardiner). Since Kay has fallen in love with Jimmy, the occasion gives the bogus maid and butler the chance to cause a rift between the host and his fiancée by performing their tasks as ineptly as possible. Shorty begins by dashing over from the kitchen yelling, "Ersters! Ersters! Ersters! Raw! Raw! Raw!," and dumping a pile of them on the table. Later he tosses the shells into a pail with a loud clanging noise. He insists that everyone have soup, which he advises should be consumed "at the height of its fever." Kay opens the champagne bottle with such a loud pop that the three diners, fearing they are being shot at, nervously raise their hands. When Jimmy tries taking a platter away from Kay because he thinks it's too heavy, he staggers and bumps against his fiancée's father, upsetting the soup onto his lap. This almost provokes a fistfight. The scene ends in total disarray when Shorty, using the tablecloth as a sack in which to hold everything left on the table, throws it wildly offstage—to the accompanying sound of crashing glass and china.

Oh, Kay! reinforced Gertrude Lawrence's position as a Gotham favorite, provided the opportunity for another winning score by the Gershwin brothers (including "Do Do Do," "Maybe," and "Someone To Watch Over Me"), and successfully launched Victor Moore in the most significant phase of his musical-comedy career.

The following season Moore was in two musicals. The first was *Allez-Oop!*, his only revue. As movie-theatre showman "Roxy" Rothafel, he appeared in a scene burlesquing stage presentations at ornate cinema palaces, and he also found a spot for "Change Your Act," though this wasn't listed in the program. Within two months, however, Moore quit the revue to appear in another Aarons and Freedley musical comedy with a score by the Gershwin brothers. It was called *Funny Face*, and it starred Fred and Adele Astaire.

Originally, Victor Moore was not in this production. Under the title of *Smarty*, with its book co-authored by Fred Thompson and humorist-critic Robert Benchley (his only effort as a librettist), the show opened its tryout tour in Philadelphia in October 1927 to a generally negative reception. Benchley withdrew from the enterprise, and Paul Gerard Smith, his successor, beefed up the comedy by adding the characters of two incompetent burglars, Herbie and Chester, played by Victor Moore and Earl Hampton. Other alterations: the substitution of Allen Kearns for Stanley Ridges as Adele's romantic interest, the replacement of eight songs (including swapping "He Loves and She Loves" for "How Long Has This Been Going On?"), and the titular switch from *Smarty* to *Funny Face*. Within six weeks a highly dubious entry had been turned into a resounding hit. It opened on Broadway November 22, 1927, as the premiere attraction of Alex Aarons and Vinton Freedley's newly built Alvin Theatre ("Al" for Alex, "Vin" for Vinton) on West 52nd Street.

In the plot, Jimmy Reeve's ward, Frankie (Adele Astaire), who is something of a fibber, wants to keep her guardian (Fred Astaire) unaware of the contents of her diary, which Jimmy has put in a wall safe. At Jimmy's twenty-fifth birthday party, Frankie induces aviator Peter Thurston (Allen Kearns) and his friend Dugsie Gibbs (William Kent) to steal the diary, but they mistakenly take a bracelet while Herbie and his partner Chester mistakenly take the diary. This sets off a madcap chase in which everyone goes off to the Canoe Inn, at Lake Wapatog, New Jersey, and ends up at the Paymore Hotel in Atlantic City. Somehow matters are reasonably well resolved by the final curtain. Among the memorable Gershwin pieces that Fred and Adele introduced—either together or with others—were "Funny Face," "High Hat," "'S Wonderful," "My One and Only," and "The Babbitt and the Bromide."

The two scenes that got the biggest laughs both involved Herbie and Dugsie. In the first, Herbie, posing as a caterer at Jimmy's party, has just poured a special recipe into a punch bowl when Dugsie, the first guest, shows up. As the two exchange rambling, casual banter, they keep dipping into the bowl until, pleasantly sloshed, they ask each other about their respective professions. Dugsie tells Herbie he etches (and Herbie asks Dugsie why he doesn't scratch himself), and Herbie admits in a matter-of-fact tone that his real occupation is not catering but safecracking. When Dugsie, without the slightest show of alarm, professes a genuine interest in Herbie's work, the pathetic thief is so pleased to talk about it that he pridefully explains the importance of having the proper tools and even confides that he plans to rob the safe in the very room in which they are standing.

The other comic scene takes place at the Canoe Inn. When Chester tells Herbie that he must kill Dugsie because he had robbed the safe before they had had a chance, Herbie, left alone on the stage, curls his lip, pulls out his gun, and tries to appear menacing by practicing barking out the line, "Stick 'em up or I'll bump you dead." Dugsie enters and Herbie apologetically explains his mission. With a catch in his voice, he asks if Dugsie plans to go anyplace in particular. No, but he does have a date for dinner. Herbie ponders the etiquette of killing someone and thus forcing him to break a social engagement. His solution: he'll come back after dinner. But that won't do because Dugsie plans to go to bed right after dinner. After some further deliberation, Herbie decides that they might as well get it over with as soon as possible. He takes out a gun and proudly shows it to Dugsie, who admires it and hands it back. Dugsie, however, warns Herbie that game wardens in that area are very strict and he'd better get a shooting license, a technicality that is circumvented when Dugsie promises that he won't tattle. Herbie takes a practice shot at a tree, then discovers that he has forgotten to load the gun, and Dugsie considerately lends him *his* gun. But Herbie has another problem. He cannot shoot him if Dugsie keeps his eyes open. Why? Because when he was a little boy, his mother had given him three kittens to drown, and he couldn't do it when he looked at their sad blue eyes. Finally, the two take up positions, and Dugsie even offers to do the counting himself. Herbie shoots and misses ("I don't see any holes"), shoots again and misses (Dugsie encouragingly tells him he's getting warmer), and shoots a third time and misses (seems that he had the gun sight set for 100 yards and Dugsie is only ten feet away). Seizing his chance, Dugsie makes a direct appeal for sympathy by telling how his four children will soon be looking down an empty street for their father who will never return. By now both men are crying. Relieved that the ordeal is over, Dugsie offers Herbie a cigar, Herbie offers Dugsie a banana, and the two exit arm in arm. Wrote Brooks Atkinson in the *Times:* "It's not the business of the act that matters; it is Moore's slow, ponderous motions, the absurd shape of his bulging head, and his sobbing voice."

Hold Everything!, which opened in New York in the fall of 1928, made it three Aarons and Freedley hits in a row for Victor Moore. But the show, which ran over 400 performances, did little to add to Moore's prestige. For this was the occasion of Bert Lahr's first major Broadway appearance, and he was unquestionably the chief comic attraction. Moore's role was largely to provide a contrasting foil to Lahr's pugnacious pug. No one even bothered to cast Moore against type; in the show he was simply a cook at a prizefighter's training camp who is Lahr's best friend.

Moore did have one rambling monologue in which he reminisced about his boyhood days when his family sat around a dining table eight by twelve ("There were twelve of us who all ate at the same table"). Victor would have made it thirteen, but his superstitious father wouldn't let him sit down; instead, he and the family dog had to stand around waiting for whatever scraps of food might be tossed their way. Frequently, there would be a fight over a bone, and in desperation Victor told his father to choose between him and the dog. "My father loved me," Moore recalled, "so he helped me pack."

As something of a compensation for Moore having so little to do in *Hold Everything!*, Aarons and Freedley planned to give him the major role in their next musical, *Me for You*. The show had a libretto by Owen Davis (the author of Moore's earlier success, *Easy Come, Easy Go)* and songs by Richard Rodgers and Lorenz Hart (including "Why Do You Suppose?" and "A Ship Without a Sail"). In the story, not only was Victor Moore again cast as a bootlegger, he was also a wealthy roué who is actually the mastermind of the entire operation. Egbert Peasley (Moore) wants his daughter Janet (Betty Starbuck) and his respectable friends (including Ray Bolger) to remain ignorant of his true occupation. When Janet becomes romantically involved with assistant district attorney Rodney Stoddard (John Hundley), Egbert tries to break up the romance by appointing his bootlegging partner, Gil Stark (Jack Whiting), to be Janet's legal guardian for a month. Results: Janet and Gil fall in love, which somehow motivates everyone to go straight.

At least that was the basic plot of *Me for You* when it tried out in Detroit in mid-September 1929. Unfortunately, Moore had a role that was simply too far removed from his lovable, innocent self, and audiences found him neither believable nor funny. *Me for You* closed after two weeks, and new writers, Paul Gerard Smith (who had performed a similar rescue job on *Funny Face*) and John McGowan, were rushed in. They did a thorough rewrite. Apart from still being concerned with bootlegging, their musical—now called *Heads Up!*—had a totally different story, though it did manage to retain the services of the five leading actors.

In its new form—which seems to have been something of a variation on *Oh, Kay!*—the show was concerned with amorous, comical, and illegal doings aboard a yacht, the *Silver Lady*, as it sails from New London to Southampton. Unknown to Mrs. Trumbell, the yacht's owner, the ship's captain uses the vessel for rum-running. A detective who suspects the truth wins the assistance of Coast Guard officer Jack Mason (Jack Whiting), who is a rival of socialite Rex Cutting (John Hundley) for the hand of Mrs. Trumbell's daughter Martha (Barbara Newberry). Also on board are Georgie (Ray Bolger) and his girl friend, Betty Boyd (Betty Starbuck, now in the secondary female lead). As for Victor Moore, he was stripped of all wealth, guile, and social position to be cast as Skippy Dugan, the cook aboard the yacht who has a prison record but now considers himself primarily an inventor.

Skippy's galley, in fact, is as crowded with his brainstorms as any gadget-filled domicile inhabited by Joe Cook. As Skippy enters, he puts his coat on a hanger which flies up to the ceiling. He pushes a button, a hat rack comes out of the wall, he puts his yachting cap on the rack, and it disappears into the wall. He walks underneath his suspended chef's hat, puts on his apron, and the hat drops on his head. He then strikes a match with sparks from an emery wheel and lights the gas stove. Eggs come tumbling down a miniature scenic railway from the ceiling. Skippy pushes a button and a loaf of bread flies out of the kitchen cabinet. He catches the bread, then turns on an electric switch which raises a saw that automatically cuts a slice. When the saw gets to the bottom of the loaf, a bell rings and it rises up and repeats the same operation. Once the bread is ready for a sandwich, Skippy pushes another button and a large piece of baloney sails out of the cabinet. After a number of attempts, he manages to get hold of the baloney and, ever so daintily, cut off a slice with a pair of scissors.

Of course, Victor Moore was given most of the wheezes, all focusing on what a dim-witted fellow he is:

When a detective asks Skippy if he knows what perjury is, he replies, "Perjury is the place where naughty people go when they die." Warning Skippy against insubordination, the captain of the yacht tells him, "I don't want any backtalk. That's mutiny. And you know what mutiny means." Skippy replies: "Oh, yes. That's a show they give in the afternoon." Referring to the time he was first sent to jail for stealing a horse, Skippy explains, "It wasn't my fault. I found a rope and I took it home. When I got home on the other end was a horse."*

Moore's luck seems to have run out the following season when he became involved in two misfortunes. The first, *Princess Charming*, found the actor providing whatever comic relief there was to the melodramatic doings of old-fashioned operetta; two months later, in *She Lived Next to the Firehouse*, he played the captain of a fire company in a weak farce set in the 1890s. Count-

* Paramount's film version of *Heads Up!* was the first movie (of two) in which Victor Moore repeated a Broadway role on the screen. Also in the cast were Charles "Buddy" Rogers and Helen Kane.

ing both shows, Moore worked only ten weeks during the Broadway season of 1930–31.

And then came *Of Thee I Sing.* The idea for the musical originated with librettists Morrie Ryskind and George S. Kaufman, and songwriters George Gershwin and Ira Gershwin, the quartet responsible for the 1930 anti-war satire, *Strike Up the Band.* Soon after the opening of the earlier show, the writers began discussing the possibility of creating a musical that, unlike their first effort together, would be uncompromising in its attitude toward the follies and foibles of American politics. The cynicism induced by the Great Depression made once-forbidden subjects fair game for ridicule, and in writing their new work the authors happily skewered such institutions as Presidential campaigns, the inaugural ceremony, the office of the Vice President, Congress, the Supreme Court, foreign affairs, and even managed to get their licks in at the expense of the Miss America Beauty Contest, marriage, and motherhood. All this was achieved with the aid of a score that—notwithstanding such hits as "Love Is Sweeping the Country" and "Who Cares?"—made the production more of an American adaptation of a closely integrated Gilbert and Sullivan comic opera than a conventional song-and-dance musical comedy.

At the age of fifty-five, Victor Moore was the obvious choice to portray the timorous, self-effacing, bewildered Vice-Presidential candidate, Alexander Throttlebottom. To play opposite him, as Presidential candidate John P. Wintergreen, the writers chose a thirty-seven-year-old musical-comedy actor, William Gaxton. Gaxton provided the perfect contrast to Moore. Sharp-featured, with glossy black hair slicked back, he was a breezy, fast-talking performer who could sing passably, dance a little, and give the illusion of inexhaustible energy. Except for movie assignments, Moore and Gaxton—or Gaxton and Moore as they were always billed—remained a team for over fourteen years.

Moore's depiction of the faceless Alexander Throttlebottom was one of Broadway's classic comic performances. He first appears in *Of Thee I Sing* as he timidly enters a hotel room where a strategy session is being held by Wintergreen's political advisers at the beginning of the campaign. Forcing a smile and a self-confident greeting, Throttlebottom is confronted by a roomful of men who haven't the foggiest idea who he is. After explaining that he is indeed the Vice-Presidential candidate, he apologetically tells the politicians that he wants to be dropped from the ticket because his mother would be upset if she ever found out what he is doing. He is told that his concern is groundless ("You'll forget it yourself in three months"), and flattered by the idea that he might succeed to the Presidency, he agrees to remain a candidate. Wintergreen strides into the room. Mistaking Throttlebottom for a waiter, he takes a drink out of his hand and orders him to get a pickle. When his running mate joins in the political discussion, the surprised Wintergreen asks his name, which even Throttlebottom has momentarily forgotten. In fact, he still doesn't even know the name of his political party. "Plenty of time for that," advises Wintergreen. "The important thing is to get elected." The scene ends with everyone in agreement that the candidates will run on the platform of Love.

The ticket of Wintergreen and Throttlebottom sweeps the country, and when next seen the Vice President is in the White House. His presence, however, is not for a cabinet meeting but because he is taking a guided tour. As he absent-mindedly observes the historic rooms and paintings, he hears a sightseer ask the guide (Ralph Riggs) a question:

SIGHTSEER: Where does the Vice President live?

GUIDE: Who?

SIGHTSEER: The Vice President. Where does he live?

GUIDE *(taking a small red book out of his pocket):* Just one moment, please. Vice Regent, Viceroy, Vice societies . . . I'm sorry, but he doesn't seem to be in here.

THROTTLEBOTTOM: I can tell you about that.

GUIDE: What?

THROTTLEBOTTOM: I know where the Vice President lives.

GUIDE: Where?

THROTTLEBOTTOM: He lives at 1448 Z Street.

GUIDE: Well, that's very interesting. He has a house there, has he?

THROTTLEBOTTOM: Well, he lives there.

GUIDE: All by himself?

THROTTLEBOTTOM: No, with the other boarders. It's an awfully good place, Mrs. Spiegelbaum's. It's a great place if you like kosher cooking.

GUIDE: Think of you knowing all that! Are you a Washingtonian?

THROTTLEBOTTOM: Well, I've been here since March 4. I came down for the inauguration, but I lost my ticket.

GUIDE: You don't say? Well! First time you've been to the White House?

THROTTLEBOTTOM: I didn't know people were allowed in.

GUIDE: You seem to know the Vice President pretty well. What kind of fellow is he?

THROTTLEBOTTOM: He's all right. He's a nice fellow when you get to know him, but nobody wants to know him.

GUIDE: What's the matter with him?

THROTTLEBOTTOM: There's nothing the matter with him. Just Vice President.

GUIDE: Well, what does he do all the time?

THROTTLEBOTTOM: He sits around in the parks, and feeds the pigeons, and takes walks, and goes to the movies. The other day he was going to join the library, but he had to give two references, so he couldn't get in.

The major complication in *Of Thee I Sing* arises when Wintergreen, after promising to wed the winner of the Miss America Beauty Contest, changes his mind and marries Mary Turner (for one thing, she can bake corn muffins). Since the contest winner turns out to be "the illegitimate daughter of an illegitimate son of an illegitimate nephew of Napoleon," this provokes a diplomatic crisis with France. The problem is settled in the final scene when Wintergreen, who is not only married but the father of twins, offers the jilted girl to Throttlebottom. ("It's in the Constitution! When the President of the United States is unable to fulfill his duties, his obligations are assumed by the Vice President.")

Most of the theatre critics greeted *Of Thee I Sing* as a major cultural event. It went on to win the Pulitzer Prize for drama, the first musical so honored, and its run of 441 performances made it the longest running of any book musical of the decade.

With a single exception, Moore and Gaxton had an enviable string of four hits in a row. Their lone failure was *Let 'Em Eat Cake*, a sequel to *Of Thee I Sing* that opened in 1933. Another creation of Ryskind, Kaufman, and the Gershwin brothers, it again offered the leading performers in the same roles they had appeared in before. This time the satirical thrusts were aimed at radicals, dictators, and the stuffy members of the Union League Club, in addition to such previous targets as the Supreme Court and foreign affairs. But the tone of sharp but good-humored jesting was replaced by heavy-handedness and an insecure approach as the musical dealt with issues that did not lend themselves readily to satirical treatment. The story traced the fortunes of the Wintergreen-Throttlebottom ticket after its defeat for re-election. Wintergreen leads a blue-shirted army (uncomfortably recalling the Nazis' brown shirts and the Fascist black shirts) in a bloodless overthrow of the government, and sets up a dictatorship of the proletariat. What particularly disturbed audiences was that a baseball game between the nine Supreme Court justices and nine debt-owing nations resulted in Throttlebottom, who was the umpire, being sentenced to the guillotine for helping the foreigners to win. Of course, he is eventually saved, but there was something jarring in the sight of the poor, pathetic Vice President with his head sticking out of the guillotine hole with the knife blade poised above him. This was satire as pointless as it was cruel. *Let 'Em Eat Cake* lasted only three months.

One producer who was keenly interested in Victor Moore's newly found eminence was Vinton Freedley. No longer a partner of Alex Aarons, Freedley had become so heavily in debt that he had to flee the country to avoid his creditors. While away, he began thinking of a musical production that would help him regain his fortunes. Moore and Gaxton would be ideal for the leads, as would his discovery, Ethel Merman. Because George and Ira Gershwin were busy writing *Porgy and Bess*, Freedley was sure that Cole Porter would be the ideal songwriter to furnish the score.* He was equally confident that veteran librettists Guy Bolton and P. G. Wodehouse would come up with an appropriately witty book that would hold everything together.

Once Freedley had settled his debts, he began putting his show into production. What he wanted from Bolton and Wodehouse was a script in which a number of oddball characters meet on board an ocean liner and the ways in which they all react to a shipwreck. The producer received the book from the authors in August 1934. Within a month, on September 8, 1934, the *S. S. Morro Castle* caught fire off Asbury Park and went down with the loss of 125 lives.

Obviously, the shipwreck idea could no longer be used. Since neither of his librettists was available for revisions, Freedley turned to his director, Howard Lindsay, to make the necessary changes. Lindsay agreed to do the rewrite but only on the condition that he have a collaborator. Someone thought of Russel Crouse, the co-author of Joe Cook's most recent show, *Hold Your Horses*, and so, out of desperation, the team of Lindsay and Crouse was born. Except for the elimination of the shipwreck and the addition of certain comedy scenes, the new script followed the general outline of the Bolton-Wodehouse story, with the leading characters being a Wall Street executive (Gaxton), an Aimee Semple MacPherson-type evangelist, Reno Sweeney, turned night-club singer (Merman), and a timid gangster named "Moon-Face" Mooney, Public Enemy Number 13, who disguises himself as a clergyman (Moore).

With his first appearance in a broad-brimmed, flat-topped hat, black suit, and clerical collar, Moore almost made it seem as if the writers had at last abandoned casting him against type by giving him a role that would be closer to his true personality. The scene is the deck of the liner where groups of people are saying their farewells before the ship sails. Into their midst comes the uncertain, apprehensive figure of the Rev. Dr. Moon, carrying a saxophone case. Suddenly, someone brushes against him and knocks open the case to reveal—at least to the audience—a submachine-gun inside. Moon quickly shuts the case before any of his fellow passengers has a chance to see its contents. New trouble looms, however, when the bogus Reverend is confronted

*Possibly Porter's most impressive achievement, the score included "I Get a Kick Out of You," "You're the Top," "Blow, Gabriel, Blow," "All Through the Night," and "Anything Goes."

by the beaming presence of a fellow clergyman, who introduces himself as Bishop Henry T. Dobson:

BISHOP: Are you on your way to the conference?

MOON: No, I think I'll go to bed early tonight.

BISHOP: I mean the Westminster Conference.

MOON: No. You see, I'm not a West Minister. I'm really more in the East . . .

BISHOP: What is your field, doctor?

MOON: Why, I'm sort of a—kind of a—sort of a missionary.

BISHOP: Missionary? Where?

MOON: 'Way out in China.

BISHOP: China!

MOON: 'Way, *'way* out in China.

BISHOP: Why, I've served in China for years.

MOON: Well, I wasn't exactly *in* China. You see, I was more . . .

BISHOP: Oh, I see. You were in Indo-China.

MOON: Yes, I was in Indoor China and you were in *Out*door China.

Much relieved when the confused Bishop takes a hasty leave, Moon is faced with another crisis when he is warned that detectives are on board looking for a gangster dressed as a preacher. What to do? Stay away from his cabin, he is warned, and mix with the passengers so that he won't be noticed. He approaches one group though his plaintive urging, "Let's get groupie," is met with cold stares. He does manage to infiltrate another group until his attempts at bromidic conversation drive the people away.

Moore's timid gangster was a tower of comic strength throughout the musical. Befriending Gaxton, he helps him stowaway to be near the girl he loves, and even confesses that he's a crook who is wanted in America. ("What do they want with another crook in America?" is Gaxton's reaction.) Moore then assists Gaxton to disguise himself by helping clip the hair from a pomeranian to serve as a beard, uses his trusty machine gun, which he calls "My little pal, Put-Put-Put," to come in first in a trap-shooting contest, and sings the cheerfully wistful bit of advice, "Be Like the Bluebird," after he has been thrown in the brig. Moore must even suffer the humiliation of listening to the contents of a cablegram from the F.B.I. ("MOONFACE MOONEY PUBLIC ENEMY NUMBER THIRTEEN NOT WANTED ENTIRELY HARMLESS WOULDN'T HURT A FLEA"), to which he reacts by wailing, "I can't understand this administration."

Victor Moore's reviews were even more enthusiastic than they had been for *Of Thee I Sing*, with such appraisals as "funnier than ever" (Robert Garland, *World-Telegram*), "the funniest man of our stage" (Richard Lockridge, *Sun*), "there's no funnier or more lovable comic on our stage" (John Mason Brown, *Post*), and "he's the quintessence of musical-comedy humor" (Brooks Atkinson, *Times*). The Broadway run of *Anything Goes* was 420 performances, just twenty-one short of *Of Thee I Sing.*

Though he was acclaimed on Broadway, Victor Moore could not ignore the tempting offers from Hollywood. Beginning in 1936, he made eight films in three years, none of them with William Gaxton. The best remembered of these were *Swing Time*, in which he played Fred Astaire's buddy, and *Make Way for Tomorrow*, in which he gave his first—and highly acclaimed—performance in a dramatic role.

Vinton Freedley beckoned Victor Moore back to Broadway in 1938 for a musical comedy written by Bella and Sam Spewack based in part on their play, *Clear All Wires*. Called *Leave It to Me!*, the new show again had a Cole Porter score and Moore's co-star was again the ebullient William Gaxton. For the first time in their partnership, however, Moore's role was clearly dominant. The actor emerged here as Alonzo P. Goodhue (known as "Stinky" because he could never stop his barber from putting smelly things on his hair), who is chosen as the United States ambassador to the U.S.S.R. only because his wife (played by Sophie Tucker) has made a sizable contribution to President Roosevelt's re-election campaign. Though Goodhue much prefers to stay home in Topeka pitching horseshoes with the boys, he reluctantly accepts the post—and then spends the rest of the time trying to get himself recalled with the aid of foreign correspondent Buckley Joyce Thomas (William Gaxton).

At Buck Thomas's instigation, Goodhue tries three different methods to achieve his goal. The first occurs when the German ambassador to Moscow insists on seeing the American ambassador. As the arrogant, goose-stepping, uniformed Nazi enters, he clicks his heels and stands at attention with his stomach protruding and his right arm raised in salute. Buck gets a sudden idea. "Kick him in the belly," he whispers to Goodhue, who quickly sizes up his snarling visitor. Stinky knits his brows, curls his lip, swings back his foot—and does exactly as he's been told. Unfortunately, he had not counted on the man wearing a corset, and the next day Goodhue is laid up with a bandaged foot. After the French and Latvian ambassadors have congratulated him on his nervy, wish-fulfilling act, and the British ambassador has straddled the issue ("Britain views your deed with pride and alarm, congratulates and condemns you"), Goodhue nervously awaits word from Secretary of State Cordell Hull. Finally the coded message arrives: "SOCK HIM AGAIN."

Stinky is miserable that his undiplomatic act has aroused such admiration, but Buck has another idea that he is sure will get the ambassador recalled.

Kicking a Nazi might be commendable, but what if Goodhue were to attempt to shoot someone? Surely, Hull would never stand for that. But the plan goes awry when, instead of aiming at the intended target, a Romanoff émigré, Goodhue manages to wound a counter-revolutionary who has come to assassinate the Russian foreign minister. This makes the ambassador a hero of the Soviet Union, and even Stalin shows up to dance a jig at the celebration that ends the first act.

Buck now realizes that he has been going about things in the wrong manner. The one surefire way to get Goodhue recalled is for him to deliberately try to do a good deed. For example, no government would ever allow anyone with a prescription for peace to remain a diplomat for long. So Buck dreams up the Goodhue Plan for World Peace which the ambassador reads to the assembled diplomats. For such a momentous occasion, he arrives wearing black satin knee britches, silk stockings, and a cutaway coat embroidered with eagles, gold leaves, and the Great Seal of the United States. He even sports an admiral's dress hat adorned with red, white, and blue plumes. Goodhue's plan is simple: if Germany wants to make war on France, France should let the soldiers march in without firing a shot and then send *its* soldiers into Germany. After a year, the soldiers of both countries get homesick, and they march back into their own countries. By this time a new generation of Franco-Germans will have been born. After twenty years, these Franco-Germans march into Russia and the Russians march into France and Germany. Now we have a generation of Franco-German-Russians. Before long, there will be no nations in Europe, the people will just be part of the United States of Europe. This, of course, is too much and Goodhue, as expected, is at last happily on his way back to Topeka.

The Cole Porter score yielded its share of attractive pieces (including "Get Out of Town" and "From Now On"), but none was so enthusiastically received as "My Heart Belongs to Daddy," which—with no plot motivation whatever—Mary Martin sang on a Siberian railroad station platform and then followed with a coy strip tease.* The reviewers showed their appreciation for just about everything in *Leave It to Me!*, but it was generally acknowledged to be primarily a Victor Moore evening. Writing in the *Post*, John Mason Brown referred to the actor not only as "a king among comedians" but as a king who "really can do no wrong." "He has the high virtue of remaining Mr. Moore," Brown continued. "This is his divine right. One would almost alter our form of government sooner than have him change. His Pickwickian figure is a welcome sight. So is the distress which his unfailing bashfulness can cause in him.

***Leave It to Me!* marked the Broadway debuts of both Miss Martin and, as a member of the dancing quartet that accompanied her, Gene Kelly.

His girth may have the portly dimensions of a gobbler, but his heart is the heart of a dove. The very gravity of his face only adds to the hilarity. For he is the happiest combination of a justice on the bench and a baby in distress." By the time the Broadway run was concluded, *Leave It to Me!* had chalked up almost three hundred performances before taking to the road.

Louisiana Purchase, which opened under the aegis of producer B. G. DeSylva in May 1940, was another political satire—and another triumph for Victor Moore and William Gaxton. It did, however, have its own distinctive features. For one thing, the talents of Irving Berlin were enlisted for the score, which included such appealing ballads as "It's a Lovely Day Tomorrow," "Fools Fall in Love," and "You're Lonely and I'm Lonely." For another, unlike their past appearances together, Moore and Gaxton were cast in adversarial roles rather than as allies. For still another, although Moore was again a mass of befuddled, childlike innocence, he was, at last, given the chance to display a native intelligence and even shrewdness that helped make him a formidable match for his enemies.

The Morrie Ryskind script was prompted by recent revelations of widespread corruption in Louisiana during the regime of the state's late political boss, Huey Long. Moore played the part of Senator Oliver P. Loganberry of New Hampshire who has come to New Orleans to investigate a shady operation known as the Louisiana Purchasing Company, and Gaxton played Jim Taylor, the company's president. Loganberry's first scene takes place in a swank French restaurant run by Jim's friend, Mme. Bordelaise (Irene Bordoni). A figure of righteous probity, the Senator wears a black, tight-fitting three-piece suit with a heavy metal watch chain dangling from his vest pockets, a shirt with a detachable celluloid collar, and a broad-brimmed "senatorial" hat. In his right hand he clutches a bulging briefcase. He looks around the room self-consciously, obviously unaccustomed to dining in so elegant an establishment. Seated at a table, he is welcomed by Mme, Bordelaise, who hands him a menu. Loganberry purses his lips and studies it intently. Somewhat embarrassed, he asks his hostess if it's written in a foreign language; when told that it's in French, he tries to assume a worldly air by mumbling, "I suppose it's pretty risqué." Mme. Bordelaise suggests a drink before dinner and Loganberry orders a glass of hot water ("That's good for you before a meal"), which shocks the proprietor. When he decides to order a ham sandwich, Mme. Bordelaise tells him she is insulted, and Loganberry, assuming that's because she runs a kosher restaurant, is only too happy to change his order to a hot pastrami on pumpernickel.

After Jim Taylor comes in and is introduced to the Senator, Loganberry offers him part of his sandwich, and Jim sizes him up as a pushover. The Senator tells him that he wants to see the books of the Louisiana Purchasing

Company, but Taylor says that all the records have been destroyed in "the fire we're having." "I mean 'we had,' " he quickly corrects himself. Loganberry, however, has already seen the certified public accountant who had worked on the firm's books and has subpoenaed a duplicate set, which is in his briefcase. He's sorry to say this, but the figures just don't add up right. Taylor, now desperate, lowers his voice and asks Loganberry, "I don't suppose there's a way you can forget about this, is there Senator?" Barely registering any emotion, Loganberry slowly replies, "No. No way that I can see." Then taking out a toothpick he sucks on it and calmly says, "I might even get you an extra ten years for asking me." The two men agree to have dinner at the resturant the following evening to discuss the matter further. As Loganberry is adding up the bill for the sandwich, Jim snatches his briefcase and hides it behind his back. As soon as the Senator leaves, he frantically tears through the bag—only to be caught in the act by the returning Loganberry. Unperturbed by what he sees, he tells Jim, "Oh, say, if you want to keep that briefcase to look over the records, it's all right because I had a lot of copies made anyhow." He leaves with a contented smile and a courtly wave of his hat.

Taylor comes up with a scheme to discredit the investigation by putting Loganberry in a compromising situation with an attractive girl. A Viennese refugee, Marina Van Linden (Vera Zorina), agrees to help, and she meets the Senator in a private room at Mme. Bordelaise's restaurant. Posing as Taylor's secretary, she tells Loganberry that Jim has been detained and will join them a bit later. Jim, in fact, joins them immediately, only posing as the head waiter and disguised with a flaring "French" mustache. He is accompanied by four of the other men under investigation, all similarly disguised as waiters, who wheel in a huge tureen of hot water for the Senator and proceed to spike it with gin. Loganberry likes it so much that within a short time he is tipsily shedding tears of homesickness for Washington. Since neither he nor Marina knows many people in New Orleans, they try to comfort each other by singing "You're Lonely and I'm Lonely" ("So why can't we be lonely together?"). Complaining of a run in her stocking, Marina puts her foot on the table and gives the Senator a good look at her shapely leg. Sympathetically, Loganberry takes off his shoe and shows her a hole in his sock. When Marina tells him that she has something in her eye, she sits on his lap, and the "waiters" surround them with flashing cameras. Loganberry, who is now pretty tight and unaware of what is going on, tells Jim that if he didn't have that mustache he'd be the spitting image of Jim Taylor. Jim pulls off the fake mustache. "You mean like this?" he asks. Loganberry giggles. "No," he says, shaking his head. "Wrong again."

Loganberry gets out of this pickle when Marina, taking pity on him, says that they're engaged. Later, when Jim gets Mme. Bordelaise to frame the Senator by sleeping in his hotel room, Loganberry sends for a justice of the peace to marry them. At the end, though, he is finally defeated when, being a politician, he cannot cross the picket line surrounding the building where he is to conduct his hearings.

Louisiana Purchase, with 444 performances, was the last and longest-running of the Moore-Gaxton blockbuster musicals. In 1941, Moore accepted Paramount's offer to make the show's film version, with Vera Zorina and Irene Bordoni also repeating their roles, but with Bob Hope replacing Gaxton. Moore made seven films in the next four years, with one of them, *The Heat's On*, the only movie in which he appeared with William Gaxton.

Moore interrupted his film work in 1942 to return to Broadway in a vaudeville-styled entertainment called *Keep 'Em Laughing*. William Gaxton was again on hand, with others on the bill including Hildegarde, Paul and Grace Hartman, Zero Mostel (his first time on Broadway), and Jack Cole. Unfortunately, Moore decided to revive "Change Your Act, or Back to the Woods," which turned out to be embarrassingly dated.* A sketch with Gaxton, about a timid man who tries to buy a handkerchief from a fast-talking salesman and ends up with a pearl-handled revolver, was better received. But *Keep 'Em Laughing* offered little nourishment to Moore and Gaxton followers.

Neither did the team's last two musicals, *Hollywood Pinafore* and *Nellie Bly*. The first, which opened in 1945, was George S. Kaufman's version of Gilbert and Sullivan's *H.M.S. Pinafore*, with Moore as Joseph Porter, the ruler of Pinafore Pictures ("I am the monarch of this joint/ All that I have to do is point"), and Gaxton as Dick Live-Eye, a pushy talent agent who wears an eye-patch over his left eye. The satire was awkward and the matching of Kaufman lyrics to Sullivan music seemed more appropriate for a college varsity show. The musical was gone in less than two months.

Nellie Bly, the following year, was around for only two weeks. Set in 1889, the musical found Moore playing a Hoboken ferryboat deckhand named Phineas T. Fogarty who tries to better the eighty-day record of Jules Verne's Phileas Fogg in traveling around the world. Spurring him on is William Gaxton, as the managing editor of the *New York Herald*, who does what he can to thwart the efforts of another world-circler, Nellie Bly, a reporter for the rival *World*. During his adventures, Fogarty gets seasick, flies in a balloon, and disguises himself as a woman in a harem, but all to little comic avail.

With three Broadway failures in a row, Victor Moore, at seventy, was convinced that the only sensible thing to do was to retire. He even turned down

*Emma Littlefield, Moore's wife who had appeared opposite the actor in the sketch, died in 1934. For this appearance, her role was played by Shirley Page, whom Moore had recently married.

an offer, in 1952, to recreate his role of Vice President Throttlebottom in a new production of *Of Thee I Sing*. The following year, however, he agreed to return to the footlights in a non-musical play, *On Borrowed Time*, by Paul Osborn. This was a revival of a fantasy about an elderly man who temporarily cheats death—in the person of Mr. Brink—by keeping him a prisoner in an apple tree because he doesn't want to leave his grandson. The play had first been presented on Broadway in 1938 with Dudley Digges in the lead, and that year Moore had played the role of Gramps in the West Coast production. In the 1953 version the actor won praise for a sensitive, moving, and restrained performance, all the more notable because nothing was retained to remind audiences of the customarily bumbling, befuddled Victor Moore. But theatregoers did not respond to a serious Victor Moore, no matter how affecting he was, and *On Borrowed Time* was withdrawn after only 78 performances.

The play's relatively brief run strengthened Moore's determination to resist all blandishments to perform again in public. From then on he was in virtual retirement until his death of a heart attack on July 23, 1962. He was eighty-six years old.

Moore did, however, weaken just once. That was in 1957, when Jean Dalrymple, who was then offering limited-run revivals of musicals at the New York City Center, persuaded him to play the starkeeper in a production of Rodgers and Hammerstein's *Carousel*. At first Moore refused because he feared risking public rejection once again. But Miss Dalrymple convinced him that the part would not only be perfect for him but would also give him a chance to end his career on a high note in a prestigious production. Thus, Moore joined Howard Keel, Barbara Cook, Pat Stanley, Marie Powers, Russell Nype, and Bambi Linn for the musical's three-week run, from September 11 to 29.

Victor Moore first appeared in *Carousel* in the third scene of the second act when he was seen sitting on top of a stepladder hanging stars and dusting them off with a feather duster. Just the sight of Victor Moore alone on the stage was enough to set off a roar throughout the cavernous City Center on opening night as the audience applauded, cheered, and whistled. In order that the play could continue, Moore was forced to step out of character briefly. Totally overcome, he faced the audience, took off his glasses, and rubbed his tear-filled eyes. Nothing more was needed to convince him of the deep affection with which he was still held. Almost five years later, after Victor Moore's death, that feeling was expressed in a eulogy written by Howard Lindsay and Russel Crouse which summed up his special appeal in one sentence: "He was someone that all the members of the audience wanted to take home with them."

Ed Wynn

NO OTHER BROADWAY CLOWN ever dominated a particular field of comedy as completely as did Ed Wynn. From 1918, when he first received solo star billing in *Sometime*, until 1942, when he made his final New York appearance in *Laugh, Town, Laugh*, Wynn not only inhabited a world of his own, he was almost always in total charge of it. An Ed Wynn show was literally an Ed Wynn show. Though the comedian often shared the spotlight with others—and it was usually *their* spotlight that he shared—he was never co-starred with anyone. Even in book musicals the action would come to a halt as Ed trotted out some goofy new invention, or interjected a totally unmotivated routine, or came up with an involved non sequitur story. His personality—and his alone—gave his shows their distinction and appeal.

It is hard to imagine that any of the productions in which he starred, despite the printed record, could have been without Ed's services as both writer and director. Yet even if we limit ourselves to only the official credits of his eleven major musicals, we still find Ed Wynn listed as author or co-author of seven and director of three—as well as being the producer of five and the composer-lyricist of three. Of all these shows, however, Wynn's six vaudeville-type "entertainments" (he never called them revues) were his most personal expressions. Onstage throughout most of the evening, he was the theatre's most disarming and amiable busybody, welcoming his audience, explaining what would and would not take place, introducing the individual acts, assisting the performers by lending a comic hand or encouraging them with comic observations, and holding the stage alone to tell punning jokes and make addle-brained observations. In some of his shows, he would bound into the aisle during intermission to mingle with the audience; on other occasions he would greet people in the lobby after the final curtain. In short, he did everything possible to make each individual in the theatre feel as if he were a friend being personally guided through a singularly lunatic, hilarious world.

Of all the buffoons who ever performed on Broadway, Ed was the one who made himself most closely resemble the traditional image of a circus clown. He was an androgynous cartoon character come to life. His long, puffy, oval-shaped face was whitened to emphasize his arched greasepainted eyebrows, his round, horn-rimmed glasses, and his expressive, popping eyes. Curving up from beneath his prominent nose was his crescent-shaped, red-lipped mouth, while peeping out from beneath an absurdly tiny hat were tufts of frizzy hair. All this gave Ed a look of constant wonderment and surprise as he gazed about the zany universe he had created. Adding to his clownlike appearance were Ed's colorful costumes. He had a collection of over eight hundred funny hats and three hundred bizarre jackets and coats that adorned his tall, pear-shaped frame. Also an important part of this comical getup were his flapping, over-sized shoes that he had bought in 1907 and which were patched so often that within a few years nothing remained of the original leather. Unlike a circus clown, Ed, of course, did not limit himself to pantomime. He lisped, and he giggled, and he talked in a high, nasal voice, as his hands fluttered incessantly in gestures of total bewilderment.

Ed Wynn was not so much a childlike adult (as was Victor Moore) as he was an adult who was still a child. His hats, his costumes, and his outlook all were part of a child's fantasy world. As he once told his son, the actor Keenan Wynn, "I never wanted to be a real person." In his comedy, Ed was always too ingenuously cheerful to ever be sly, or malicious, or even witty. He boasted that he never told an off-color joke; it's also true that few of his gags were intrinsically funny. It was Ed's delight in them and his complete conviction in telling them that made people laugh at his jokes (in much the same way that Jimmy Durante's delight and conviction made people laugh at *his* jokes). Part of the audience's enjoyment in Ed Wynn was in watching him enjoying himself; he was comical for the sheer joy of being comical, not to delineate a character in a story or to express an attitude toward society. Though in real life he was often morose and withdrawn ("Maybe I used up all my happiness on the stage," he once said), as soon as he put on greasepaint and donned an outlandish costume he suddenly became a warm and ebullient person who would do anything to win laughs and applause. Throughout his entire stage career, he was at his happiest making a perfect fool of himself, a designation that he was also happy to appropriate as his own self-description.

The son of a middle-class Philadelphia hat manufacturer, Ed Wynn was born Isaiah Edwin Leopold on November 9, 1886. His father, who had been born in Prague, was a hard-headed businessman, and young Ed sought escape from his dominance by retreating into the fantasy world of the theatre. He could never remember a time that he wanted to do anything but make people laugh. This became particularly irritating to the elder Leopold when his son would try on the latest millinery creations for the amusement of customers.

Ed Wynn accompanying a diva in the *Ed Wynn Carnival* (1920), the first show the comedian produced himself. *

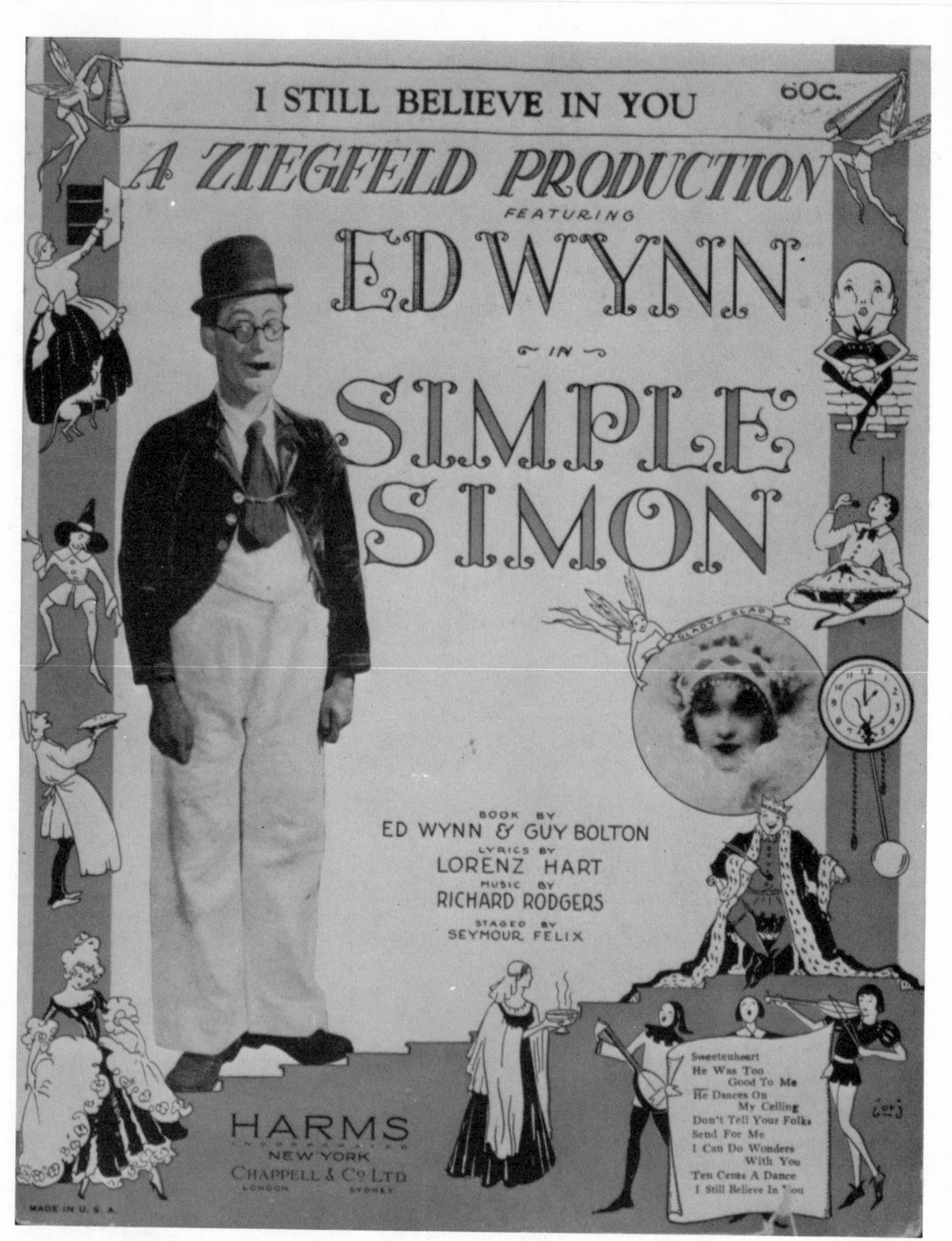

The sheet music cover of Ed Wynn's 1930 book musical, *Simple Simon*.

Alan Edwards, Doree Leslie, and Ed Wynn in *Simple Simon*.*

The sheet music cover of the 1937 anti-war musical, *Hooray for What!* (Note that the names of Jack Whiting and Vincente Minnelli are misspelled.)

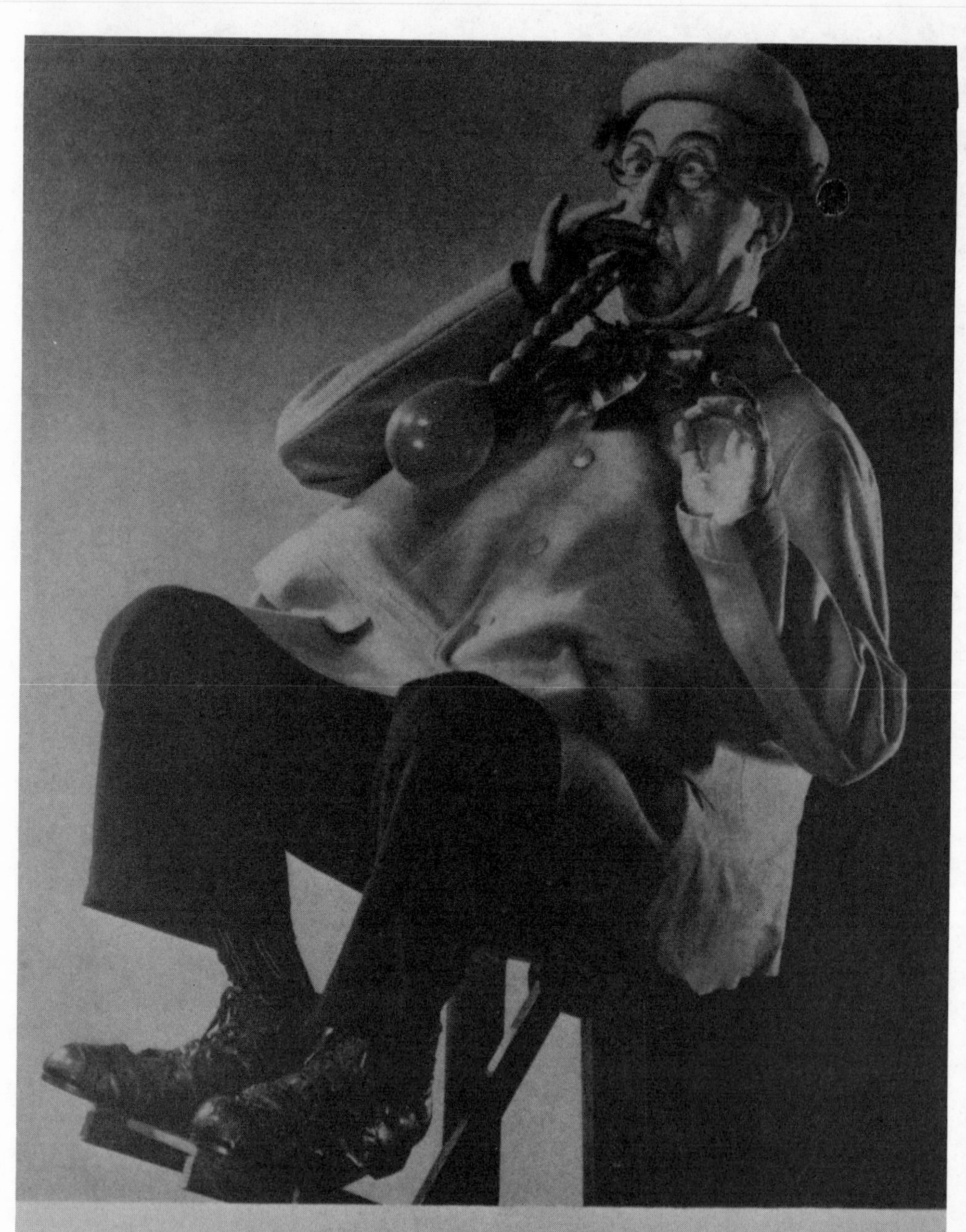

The Playbill cover of Ed Wynn's *Boys and Girls Together* (1940). (Playbill ® is a registered trademark of Playbill Incorporated, N.Y.C. Used by permission.)

Because his father was firmly opposed to Ed's having anything to do with the theatre, life became so miserable at home that at the age of fifteen the youth ran away to join a traveling repertory company. After being stranded in Bangor, Maine, he returned home, and his father, hoping to make him give up his wild idea of becoming a comedian, put him to work in his business. But as soon as Ed had earned enough money, he ran away again, this time to New York. It was then that he adopted his stage name by splitting his middle name in two. He teamed with another young comedian, Jack Lewis, in an act called Win and Lose, which played small-time vaudeville engagements for two years. In one of their routines, "The Freshman and the Sophomore," Ed kidded the collegiate life by coming on stage wearing a wide-brimmed floppy Panama hat, smoking a meerschaum pipe, and carrying a bulldog in his arm. His opening lines were "Rah! Rah! Rah! Who pays the bills? Ma and Pa!" After becoming a single in vaudeville, Ed introduced an act called "The Boy with the Funny Hats." As something of a holdover from his early days clowning in his father's showroom—and as an indication of his lifelong delight in goofy headgear—he got laughs by twisting his Panama hat into twenty-eight different shapes.

Ed Wynn was not quite twenty-four when he made his Broadway debut in 1910 in a long-forgotten musical called *The Deacon and the Lady.* Harry Kelly starred as a deacon who owns land in Montana which he sells to a young couple, and Wynn played his assistant. Ed wrote his own material and did his hat-twisting act, but the show was off the boards in two weeks.

Back in vaudeville, he introduced a new act, "The King's Jester," in which he had to make a gloomy king laugh or be beheaded. While the audience out front may have been convulsed by Ed's quips, the king remained sullen-faced as Ed frantically tried to avoid the executioner's blade. Finally, in desperation, he called the king to one side and whispered in his ear. With that, the monarch broke into uncontrollable laughter. Ed closed the act by exclaiming, "I didn't know you wanted to hear *that* kind of a joke!"—and ended the scene by kicking the king in the seat of his pants.

Ed repeated this act as part of the first bill that played the Palace on March 24, 1913. Possibly because he was a last-minute substitution, Wynn received billing far down on the program. The other attractions on that first bill were Eight London Palace Girls (dancers), McIntyre and Harty (comedians), a condensed version of Leo Fall's operetta, *The Eternal Waltz,* the Four Vannis (a high-wire act), La Napier Kowska (an interpretive dancer), Ota Gyri (a violinist), Hy Mayer (a newspaper cartoonist), and a one-act play by George Ade.

Producer Florenz Ziegfeld was sufficiently impressed with his appearance at the Palace to sign Wynn for the *Ziegfeld Follies of 1914,* in which he shared the comedy assignments with Leon Errol and Bert Williams. The eighth annual edition of the revue opened at the New Amsterdam Theatre on June 1, and ran for 112 performances. As described in the unsigned *New York Times* review, "The book, what little there is that can be called a book, is dangerously close to the borderline of stupidity, but there are girls and girls and girls and plenty of real comedians and many more real dancers. and when the show begins to lag, somebody dances or tells a story, or the Ziegfeld chorus comes skipping on, and all is forgiven." Ed Wynn—as "Joe King the Joke King"—did his "King's Jester" act as part of the first scene in the show, the Reception Hall in Hell, which was presided over by a female Satan (Vera Michelena), surrounded by scantily clad chorus devils. Wynn also appeared as a tango instructor in a "Tango Palace" sequence with Leon Errol as an inept pupil and Ann Pennington as a high-kicking visitor.

Wynn's *Follies* debut may have been less than sensational, but Ziegfeld signed the comedian for a repeat engagement the following year. In addition to holdovers Errol and Pennington, the 1915 *Follies* also featured W. C. Fields, Ina Claire, Mae Murray, and a young hoofer named George White. Ed Wynn scored best in a movie rehearsal scene in which, seated in the audience, he played a director giving instructions to young lovers Mae Murray and Bernard Granville and villain Leon Errol as they appeared on a screen suspended over the stage.

Wynn's penchant for horning in on other performers' acts—later to become a formalized part of his comic routines— was first apparent when the *Follies* played Boston after its Broadway run. During W. C. Fields's pantomime pool-table specialty, Fields became aware that the laughs were coming in the wrong places. The reason, he soon discovered, was that Ed was under the table reacting with clownish grimaces and gestures. Fields proceeded with the act as if nothing were amiss, but once Ed popped his head up, he clouted him so severely with his cue that it nearly knocked Wynn unconscious. Though Ed apologized for his unprofessional behavior it was a long time before Fields spoke to him again.

The following year, Ed signed with the Shubert brothers for his next three Broadway appearances. This turned out to be a fortunate step since these shows offered little competition from other comedians, and the brothers let him do pretty much what he wanted. In fact, with his very first Shubert revue, *The Passing Show of 1916*, Ed Wynn was unquestionably the leading attraction. He was first seen midway through Act One when a spotlight picked him out seated in a box as a member of the audience. Turning to the man next to him, Ed loudly proclaimed, "This show is terrible so far." The man agreed, saying that what was most needed was a good comedian. "Yes," Ed concurred, "someone funny like Ed Wynn." "Why even you'd be funnier," the man said. Overhearing the conversation, the show's manager, also in the audience, then offered Wynn a part in the revue if he thought he could do any better than the actors on the stage. "Am I funnier than they are?" Wynn giggled. "Don't make me laugh."

Ed's main sketch in the show was called "The Modern Garage," in which he appeared as an auto mechanic puttering around with much hammering and many explosive noises. Though the United States had not yet entered World War I, there was a severe fuel shortage at the time. Audiences particularly appreciated the bit when, at the request of a customer, Ed opened his safe, took out a small bottle of gasoline, and sold the man a few precious drops. Later in the scene, mistaking a blind man for a prospective automobile buyer, he put the man in a car, then turned on an electric fan so that he would think the wind was blowing in his face as he rode through the country.

The titles of Ed's next two revues, *Doing Our Bit* and *Over the Top*, both offered in the fall of 1917, indicated the theatre's increasing concern with the war effort after the United States had entered the conflict. Ed's comedy, however, seldom strayed beyond his own private world. *Doing Our Bit*, which played at the Winter Garden, also marked the first occasion that the comedian appeared at intervals during the program to prattle aimlessly before the curtain while the scenery was being changed.

In his first scene, Ed showed up as a Guest at a fashionable banquet. Attired in formal dress, the Guest enters the room carrying a bouquet of flowers. Blocking his path, however, is a hatcheck boy. At first the Guest politely declines the offer to check his hat, but the attendant is persistent and the man is unable to get past. So determined is the attendant, in fact, to rid the Guest of his hat that he even tries pulling it off his head. A scuffle ensues, and both men are knocked to the floor. Finally, in desperation, the Guest takes out a gun, shoots his assailant, and deposits the floral bouquet over the body. This drastic remedy wins the hearty approval of the other banqueters, and the Guest is called upon to make a speech. "The best after-dinner speech is 'Put it all on one check,' " he says sagely, and sits down.

Later in the revue, Ed appeared as a barker at a carnival midway. "Here you are, ladies and gentlemen," he cries out. "The greatest collection of freaks on earth. The big sideshow. Here is the thinnest fat man you ever saw and the biggest midget in the world. Here is the tattooed lady, Dottie Marks. Come look at the pictures. The lady at one time had a sweetheart tattooed on her arm and several weeks later found that she had a husband on her hands." Then raising a curtain to reveal a large skeleton and a small skeleton, the barker points to them explaining, "Here is Napoleon and here is Napoleon as a little boy."

Ed was taken out of *Doing Our Bit* after two months to join the cast of *Over the Top*, a scaled-down Winter Garden revue then playing at the 44th Street Roof Theatre. The show, which featured Justine Johnstone and served to introduce Broadway to the talents of Fred and Adele Astaire, had already opened with another comic, T. Roy Barnes. After it had been running only a month, Wynn was rushed in to replace Barnes.

Ed's opening monologue, in much the same oddly egotistical vein as his opening bit in *The Passing Show*, was a rambling explanation of how he happened to be there. Following a chorus number, Ed came out waving to the departing dancers. "How do you do, girls?," he greeted them. "Get some clothes on while I talk to the audience . . . You remember me, Ed Wynn, the big comedian? The reason I asked if you remembered me was becaue it must seem funny to see me in a show of this kind. I've always been in big Winter Garden shows and the *Ziegfeld Follies*. In this show I am the plot. How I came to join this show is really very interesting. When Mr. Shubert first produced this play he found that he had everything he needed but the plot. He said, 'Ah!'—I don't really know whether he actually said 'Ah!' or not, but it sounded like 'Ah!' I was standing way back. 'Ah!,' he said, 'I have men for the songs, girls for the girls, and Wynn for the plot.' Well, I'm very good as the plot. You will notice that as I run through the show I make it very interesting. I unravel situations and complications. In fact, the show depends a lot on the plot. You will notice the difference between my lines and the author's. Mine are very good . . ."

Though *Sometime*, in 1918, turned out to be one of Ed Wynn's biggest hits—as well as his first starring vehicle—his association with it was something of an accident. That summer, producer Arthur Hammerstein had tried out the Rudolf Friml–Rida Johnson Young musical in Atlantic City, but it was soon shut down for repairs. Wynn agreed to take over the leading comedy role (originally played by Herbert Corthell), but only on the condition that he be allowed to rewrite the book and contribute some new song lyrics. Opening on Broadway in October, *Sometime* (a rubric intended to suggest its closeness to Mrs. Young's still-running hit *Maytime)* told a backstage romance of a leading lady (played by Francine Larrimore) who, on opening night, relates via flashback the story of her love affair with a handsome man seated out front. Wynn played Loney Bright, the show's property man and owner of a theatrical boarding house. The jokes were all vintage Ed Wynn:

"Why is six o'clock like a pig's tail?"
"Twirly."

"What is a man to do who can't make both ends meet?"
"Make one end vegetable."

"Did you give me a dirty look?"
"I didn't give you a dirty look. You've got one but I didn't give it to you."

Loney even has a girl friend, Mayme Dean (played by twenty-five-year-old Mae West), a chorus girl for whom he has written a play, and which he describes to her: "The play is called *The Moonshiner's Daughter*, and your father is making whiskey for his own private use. He has about a million barrels. That's why I'm in love with you. As the curtain goes up, you are feeding a lot of mules. There are revenue officers all around. The revenue men go over to the mules and pick you out. You've got a hat on. They say to you, 'Someone has been making whiskey around here. Do you know where there is a private still?' And you say, 'There'—pointing to me. You say, 'There. He could have been a lieutenant but he didn't get the commission so he's a private still.' "

The Shubert Gaieties of 1919 was Wynn's last revue for the brothers. The best-received routine concerned Ed's laughter drive—patterned along the lines of a war-bond drive—in which his goal was to register 1000 laughs on a huge meter. In one laughter-provoking attempt, Ed inquired if Flo Ziegfeld were present because if he laughed during a Shubert show that alone would be worth at least 1000 laughs. In addition, there were three occasions in which Ed delivered his casual monologues in front of the curtain.

The Shubert Gaieties had one special distinction apart from what occurred onstage. Within a month after it had opened on Broadway, the newly formed Actors' Equity Association called a walkout against the powerful Producing Managers' Association to protest salaries and working conditions. Though he had no personal grievances (his weekly salary of $1750 was among the highest in the theatre) and he was not then a member of Equity, Ed became one of the strike leaders whose efforts helped force the producers to accept the union's demands. Since managers still retained the right to hire and fire and since the Shuberts were incensed at Ed's involvement in the strike, the comedian was not rehired when the *Gaieties* resumed its run. Nor would any other member of the producers' association sign him either. This left only one course of action: in partnership with a producer from Detroit, B. C. Whitney, Wynn decided to sponsor his own show, the *Ed Wynn Carnival*. When word of this got around, rival managers then threatened to blacklist any actors or even sognwriters who had anything to do with the new production. Wynn's solution was simply to enlist his cast from the ranks of vaudeville and the circus and, since he had had some experience writing songs for his vaudeville act, to supply his own music and lyrics.

Thus, through necessity, Ed Wynn was able to create the kind of proprietary entertainment with which he was most closely identified during his career. It was, of course, entirely congenial to his personality, allowing him to open up the stage even more than in previous shows to a world of his own madcap creation. Now totally in his element, he was the irrepressible hopping, skipping, joking host of the entire evening.

The *Ed Wynn Carnival* opened at the New Amsterdam Theatre on April 5, 1920. Ed's first appearance in the show prompted a five-minute demonstration from Actors' Equity members who had packed the house, but the reviews the next day confirmed that Ed would have scored a resounding hit even without a cheering section. As Charles Darnton of the *Evening World* commented, "He is always busy at something, from explaining the plot—of which there isn't any—to helping the acrobats. Popping out here and there unexpectedly, he is like a jack-in-the-box working overtime. But if he feels a heavy responsibility resting on him, he doesn't for a moment let you feel it. No matter what he says or does is absurdly funny. His good nature and energy are equally remarkable." Among Ed's routines and stunts were his comical introductions to the various acts, his ball-juggling (which he prefaced by requesting orchestra leader Max Steiner to play something in a jugular vein), his inept violin playing, his specialty as a human lightning calculator, his use of an open piano as a huge ash tray, and his appearance in the lobby after the final curtain bidding his homeward-bound guests goodnight.

Wynn continued this personal, intimate style "entertainment" in *The Perfect Fool*, a descriptive title he appropriated after overhearing a woman at the New Amsterdam box office requesting tickets for her parents who were so anxious to see Ed because "they think he's a perfect fool." The new show featured a mind-reading act in which members of the audience would whisper inquiries to Ed's assistants in the aisles, which the blindfolded comedian would then answer from the stage. (When one first-nighter, the *Tribune*'s critic, Percy Hammond, asked a confederate, "Where is the bottle of scotch Ed promised me?," Wynn instantly answered, "Tell Mr. Hammond that he will find it in the box office as he goes out.") *The Perfect Fool* was also the first show in which Ed introduced his wacky inventions, such as the odorless cheese fork, the silent soup spoon, the coffee cup with a hole in it for easy pouring into the saucer, and the typewriter whose moving carriage is equipped with corn on the cob so that the diner need not move his head.* As expected, the comedian introduced performers with little relevance to what they would be doing, he again became involved with acrobats, and he repeated the business of chatting with fans in the lobby following the show. After playing for 275 performances on Broadway, the Perfect Fool took *The Perfect Fool* on the road for two years.

*One celebrated invention, the eleven-foot pole for people you wouldn't touch with a ten-foot pole, was probably introduced in vaudeville or on the radio since it cannot be traced to a Broadway show.

For his next production, in 1924, Ed used a program billing that he would repeat for his three subsequent "entertainments":

ED WYNN
(The Perfect Fool)
presents Himself
in
"THE GRAB BAG"

Though the show offered few surprises, nobody seemed to mind. Using the same formula as in the *Ed Wynn Carnival* and *The Perfect Fool,* Wynn even repeated some of the gags and routines from these shows (including his corn-on-the-cob typewriter and his involvement with acrobats). But for most theatregoers the important news was that Ed Wynn was back, regaling the audience with his lengthy, pointless anecdotes, telling his jokes ("The difference between a canoe and a Scotchman is that the canoe tips"), and announcing that a Russian octet would sing folk songs titled "He Eats French Dressing Every Night So He Can Wake Up Oily in the Morning" and "She May Have Been a Schoolteacher but She Never Had Any Class."

The plotless Ed Wynn show came to a temporary halt in 1927 when revue impresario George White prevailed on the comedian to star in White's first book musical, *Manhattan Mary* ("Clean from Beginning to End," as the reassuring program note advised). The producer's self-serving tale dealt with the way chorus girl Mary Brennan (Ona Munson) becomes the leading lady of the *George White's Scandals* with the help of Crickets (Ed Wynn), who is both cook and waiter at a Greenwich Village lunchroom owned by Mary's mother. Crickets is, not surprisingly, an inventor (he has devised a pair of windshield wipers to attach to eyeglasses whenever a person eats grapefruit), and he is also extremely obliging. When a tough-looking gangster comes in roaring, "I'm so hungry I could eat a horse," Crickets simply rushes through the kitchen door and comes back with a real horse. After the audience laughter has died down, he tops the sight gag by asking, "Ketchup or mustard?"

In order to help Mary take over the *Scandals'* leading role when the star, Helen King, has come down with pneumonia, Crickets accompanies her to the theatre where understudy auditions are being held by George White himself. For no apparent reason, Crickets is absentmindedly carrying a rope. When White asks him where he got it, he replies, "I don't know where I got it. I didn't even know I had it. I either found a piece of rope or I lost a horse."

Since Crickets is not above giving advice, he suggests to White that he use the "egg-wiped" method of picking chorus girls for understudies. White never heard of it:

CRICKETS *(exasperated):* I'm telling you what it is. Egg-wiped! Egg-wiped! They spell it E-G-Y-P-T.

WHITE: Oh, Egypt.

CRICKETS: That's what I mean. That's theatrical history. It goes back to 250 BC when the Egyptians used to pick chorus girls for understudies for their musical comedies.

WHITE: Why, there weren't any musical comedies then.

CRICKETS: Yes, there were. What are you talking about? I still use some of the jokes. What was the name of the Egyptian king at the time? I read his name recently. Not Pyorrhea. I have it. Tootie Ankerman. It's an historical fact that in 250 BC Tootie Ankerman put on a musical comedy but the leading lady became suddenly ill. That's the same predicament that you're in. The afternoon before the opening peformance she was rehearsing a flute solo for the second act, but the first finger of her left hand got stuck in one of the holes. She had just blown into the flute but the note didn't come out. The air pressed against her finger and it began to get swollen. Didn't you ever read that story? There she was with a flute dangling from her finger. It was very painful so she went to a drug store for some iodine, but they didn't have any. All they had were sandwiches. As she came out of the drug store, she stumbled and the flute came loose. But she still couldn't go on. King Tootie Ankerman didn't have anyone to replace her. They didn't know what understudies were in those days, so he just lined up the girls and that was the beginning of this method. He simply counted seven from his left and he took that girl.

Though skeptical, White agrees to give the method a try. He lines the girls up, but Mary, who had, of course, been the seventh girl from the left, has been moved to the eleventh spot. Unfazed, Crickets tells White that the African method, also very good, uses the eleventh girl. Thereupon White auditions Mary who dances a few steps and nervously stops. "That's enough! That's enough! I'm satisfied!," exclaims Crickets.

Mary takes over the lead, and the company goes into a rehearsal for the high-stepping "Five Step" number, which actually serves as the first-act finale of *Manhattan Mary.* White even allows Crickets to conduct the show's orchestra, thus offering the tradition-defying sight of the curtain coming down with the star of the musical waving his arms violently in the orchestra pit instead of occupying center stage.

Though scoring a hit in the *Scandals,* Mary has a fight with her boy friend who is against her stage career. She leaves the show to star in the *Folies Bergère* in Paris, but returns a year later for a reconciliation. Crickets has arranged a welcome-home party which he expects Mayor Walker to attend. When Bob, the stage manager (played by Harland Dixon), asks if he knows the mayor, Crickets replies that he doesn't exactly *know* the mayor but he *could* know the mayor:

CRICKETS: My uncle was once mayor of New York, you know.

BOB: He was?

CRICKETS: Well, he wasn't exactly mayor, but he was nominated for mayor.

BOB: He was?

CRICKETS: I shouldn't say he was nominated, but if his name had been called out at the convention he would have been nominated. But he didn't go to the convention. His desk is in the mayor's office.

BOB: It is?

CRICKETS: It's not exactly in the mayor's office, but my uncle does have a great outside job.

BOB: A great outside job?

CRICKETS: He's a street cleaner. That is, he would have been a street cleaner but he didn't get the job. My aunt has been a great help to my uncle all during his career. My aunt is an authoress, you know.

BOB: An authoress?

CRICKETS: Well, she owns a book store. It's the same thing.

BOB: It is?

CRICKETS: Well, she doesn't own it, she works in it. She and my uncle have been happily married for twenty years.

BOB: For twenty years?

CRICKETS: Well, ten years. As a matter of fact, I don't know if they're even married at all.

Trying to analyze what made Ed Wynn so funny, critic Robert Benchley, somewhat at a loss for words, admitted in *Life Magazine:* "All the writing in the world never seems quite to explain the phenomenon . . . It is, we feel, that detached, wistful acceptance of the world and its burdens which shines through those spectacles which is . . ." Making another effort, Benchley concluded his review by noting "that inexorable sangfroid even in the face of what must be intense personal excitement, that understanding of the ages, that . . ."*

Early in 1930, Ed was back on Broadway in another musical comedy with a tenuous plot. *Simple Simon* was its name, and this time the sponsor was Florenz Ziegfeld (with Ed as the uncredited co-producer) and the songs were written by Richard Rodgers and Lorenz Hart. Since Ed's comedy always reflected a child's view of the world, it seemed only fitting that he should be involved—at least once—in a story consisting of fairy-tale characters. Wynn and Guy Bolton put together a script in which Ed played Simon Eyyes, the owner of a Coney Island newsstand who dreams of Cinderella and Prince Charming and their adventures in two rival kingdoms, Dullna and Gayleria.

Ed, of course, made sure that the plot was never allowed to get in the way of his gags, inventions, and specialties. Prohibition and the Depression even gave him a chance for topical humor. "I always like reading fairy tales," Simon tells a customer at his newsstand. "Do you know the one about the man who couldn't get a drink in New York? That's a wonderful fairy tale." When another customer informs him that business is looking up, Simon replies, "It has to. It's flat on its back."

A dream sequence in a drug store finds Simon working behind the soda fountain. "A banana split," orders a customer. Simon is shocked and wants to know what they fought about. What did *who* fight about? "Abe and Anna." In the same scene, Ruth Etting shows up to audition a song, and Simon offers to accompany her. "Can you play in A-flat?," she asks. "I can play in any flat if I have the key," he answers. Simon then introduces his latest invention, a piano fitted onto a tricycle. With Miss Etting perched on the piano singing the sentiments of "Ten Cents a Dance," he accompanies her while wheeling the contraption around the stage.

Among Ed's other inventions introduced in *Simple Simon* were a nightgown that won't wrinkle because it only has sleeves, a hair tonic that can be poured on a grapefruit so that it may be sold as a cocoanut, a cigarette lighter with a box of matches attached because no lighter ever works, and a mousetrap sealed shut so that the mice, frustrated at being unable to get at the bacon, scurry out to find another house.

Unquestionably—at least for those who were there—Ed's most memorable line in *Simple Simon* was spoken in a setting made to resemble a woodland clearing. All Ed did was to come out carrying a duck and a ham in a picnic basket, spread a tablecloth on the ground, and exclaim with boundless enthusiasm, "I love the woods! I love the woods!," and audiences were con-

*Ed Wynn repeated his role of Crickets in Paramount's 1930 film version of *Manhattan Mary*, retitled *Follow the Leader*. This was Wynn's only stage role that he also played in the movies. Others in the cast were Ginger Rogers (as Mary Brennan) and Ethel Merman (as Helen King).

vulsed. And they continued to be convulsed every time he uttered the same words.

The following year, the comedian returned to Broadway in the more accustomed surroundings a freewheeling, plotless, self-presenting "entertainment." Titled *The Laugh Parade,* it was as much of an all-Ed Wynn show as previous similar outings except that he now left songwriting to the more experienced hands of Harry Warren, Mort Dixon, and Joe Young. There were nineteen scenes in the production and Ed was in almost every one of them. He began the evening by explaining how his show was different from other theatrical fare, and by introducing his cast members as they emerged from a cage.

The Laugh Parade then proceeded with a proper mixture of the expected and the unexpected. Ed invented a brown derby with a removable brim so that the wearer could tip his hat in winter without catching cold. He demonstrated his ability as a quick-change artist by impersonating Fagin in *Oliver Twist, The Hunchback of Notre Dame,* and a vaudeville juggler in green satin tights. He boasted of his parrot's pedigree: "It's one of the Parrots of Wimpole Street." Explaining that he had just come from an air-cooled movie theatre, he appeared in a raccoon coat and hat to explain the film's plot, which had something to do with a man who has an apartment on the 99th floor of the Empire State Building but who also has a place on the fourth floor where he can stay when he doesn't want to go home nights.

Ed, of course, spent the evening popping in and out of other people's acts. Alarmed at what was happening to the female member of an Apache dance team, he upbraided the male member with, "You wouldn't do that if there were a man around." He dashed on stage to catch one of drummer Jack Powell's sticks as it flew through the air. He accompanied singer Jeanne Aubert by playing and pedaling his piano on wheels. In addition, he found time to operate a one-man Punch and Judy show, lead the pit orchestra, and ride a camel. He even inaugurated his vocal trademark of punctuating a line with the high-pitched, echoing sound of "So-o-o-o." (An expression used by his mother, it was first tried out as a private joke when she was in the audience during the Philadelphia tryout.)

All the theatre critics paid due homage to Ed's inimitable fooling. "He is the only complete master of his own special brand of nonsense," wrote John Mason Brown in the *Post.* "He has no imitators who can succeed at imitating him. He is a comic law unto himself. What he does begins and ends with him. Mr. Wynn succeeds in wiping out all recollections of the everyday world outside the theatre, and in making the laughter of the moment more important than life itself. He is the king of nonsense and the emperor of idiocy."

The Laugh Parade, in addition to being one of Wynn's greatest successes, also indicated the new power of radio as an entertainment medium. During most of the show's run, the Tuesday evening performance had to be cancelled so that the star could make his weekly broadcast as Texaco's Fire Chief.

The heightened awareness during the 1930s of the danger of another world war even extended to the traditionally frivolous area of musical comedy. Three of the decade's productions—*Strike Up the Band*, *Leave It to Me!*, and *Hooray for What!*—offered not only glittering, song-filled evenings but also took satirical views of a possible future conflict involving the major powers. In order to help make them commercially palatable, they all starred notable theatre clowns: the first two had Bobby Clark and Victor Moore, the third had Ed Wynn.

Hooray for What!, which opened on Broadway late in 1937, found Ed in his customary role of inventor. Other than that, however, the Howard Lindsay–Russel Crouse–E. Y. Harburg script compelled the most clownlike of clowns to face the terrifying real world of poison gases, munitions-making, diplomatic double-dealing, espionage, and actual warfare. It did not provide the most congenial surroundings for the Perfect Fool—even though it did give him the chance to make smoke come out of his ears while he was experimenting in his laboratory—and Ed was unable to resist including, without relevance to the plot, such vaudeville specialties as The Briants, a slow-motion tramp act featuring a contortionist with a collapsible body, and Al Gordon's Dogs, who spent most of their time trotting in a circle around Ed ("I never felt so much like a tree in all my life").

In the story, Ed played Chuckles, whose inventions include crossing a silkworm with a moth to create lace curtains, and a gas to make cows lie on their backs at night so that the cream comes out on top in the morning. Chuckles's most successful brainchild, however, is a gas to kill worms in his apple orchard which, it is soon discovered, can also kill human beings. The inventor is hired by the Harriman Munitions Works and is made a delegate to an international peace conference in Geneva. There it doesn't take long before all the peace-loving nations are after his formula. At the Hotel de l'Espionage, a lady does manage to steal the plans, but, with the aid of a hand mirror, she copies them down backwards and they turn out to be Chuckles's formula for his harmless brotherly-love laughing gas.

As a member of the peace conference, Chuckles takes the opportunity to lecture his fellow delegates. "Do you know what's wrong with the world?," he asks. "England is in Palestine, Italy is in Ethiopia, Russia is in Spain, Germany is in Austria, Japan is in China. Nobody stays home." He also takes up such weighty matters as the war debt ("Do you fellows know that if you miss two more payments, America will own the last war outright?"), civil liberties ("We in America realize that free speech is not dead in Russia—only the speakers"), and his brotherly-love gas which will bring peace to the world (this so enrages the diplomats that four of them challenge Chuckles to a duel). Eventually, after war has been declared, the laughing gas does manage to laugh war off the face of the earth (at least at the Winter Garden).

Prompted by the show's title, reviewers enthusiastically proclaimed their hoorays for Ed Wynn. As Sidney Whipple observed in the *World-Telegram*, "He is on the stage almost constantly, this superclown with his manifold inventions, his unlimited wardrobe and his almost copyrighted method of enforcing his own jokes by his own paroxysms of mirth. The formula is the same, but it has been toned down a trifle. The giggle is a little less insane and the jokes—and their manner of delivery—much, much better."

Soon after the close of *Hooray for What!* in May 1938, Wynn suffered a number of personal blows. His divorce from his first wife, his brief, unhappy second marriage, and his protracted, financially draining problems with the I.R.S. resulted in a breakdown and a period of depression. In the fall of 1940, however, Ed bounced back on Broadway as if nothing had happened to present himself in his latest carnival–grab bag–laugh parade, *Boys and Girls Together*.

Ed opened his evening of falderal by scampering out of a theatre trunk with mothballs spilling all about him. He was there, as usual, to explain how unusual his show was. To illustrate, he solemnly proclaimed that his would be free from the customary boy-and-girl chorus number that opens most musicals—and then proceeded to offer a customary boy-and-girl chorus number as the kind of thing people would *not* see in his production. He was also pleased to advise that all of the performers in his show were recruited exclusively from Mississippi showboats. "I bred my cast on the waters," he announced with a chuckle.

Predictably, Ed Wynn was onstage almost the entire evening. As a harried traffic policeman (his domain is bounded by 284th street on one side and the Catskills on the other), he tries to disengage two acrobats whose arms and legs have become entangled. Once he gets involved, however, Ed discovers that he is not only unable to pull the acrobats apart but that he is also unable to extricate himself. With his face registering confusion, determination, and panic, everytime he swings his arms or legs he becomes even more tightly knotted than before.

In a woodland scene, Ed appears in an incongruous Chinese mandarin outfit rowing a boat with a double-bladed paddle. After carefully examining a road map, he throws an anchor overboard. He is there, he advises, to do a little duck shooting. Demonstrating his skill as a marksman, he takes out a gun with a curved barrel and announces that he will shoot his target backwards with the aid of a mirror held in his left hand. With gun poised, he slowly brings the mirror to a point where he catches sight of his own reflection. "Oh, for goodness sakes!," he squeals, shuddering with horror, then giggles when he realizes that he is only looking at himself.

To introduce his eight beautiful "Cocktail Hour Girls" properly, Ed presents each one in turn as she walks languorously down a flight of stairs at center stage. First, however, the stairs must be carpeted, and Ed obligingly does the job himself. With head down and his body resting on the upper steps, he installs a red carpet working his way on his elbows and knees from the top to the bottom. Pausing during his labors, he looks at the audience and asks wistfully, "Does Katharine Cornell do this?"

Riding his piano-tricycle, Ed brings out featured singer Jane Pickens perched atop the piano. Ed offers to be her accompanist. "I didn't know you could play," Jane tells him. "I couldn't until I hurt my hand about three months ago," Ed replies. "I asked the doctor if I could play the piano after it healed. He said I could and after he took the bandages off I could play."

One of the show's high spots—though it was based on a previous Ed Wynn routine—came during a juggling act. As the Six Willys furiously fill the air with whirling Indian clubs, Ed walks on stage with a purposeful look carrying a stepladder. He stops, opens the ladder, and climbs to the top at the very moment that a club comes flying in his direction. Ed catches it, descends the ladder, folds it, and hurries off into the wings.

According to the unrestrained Broadway critics, *Boys and Girls Together* was the show in which Ed Wynn was "a classic zany" (Lockridge, *Sun*), "a magnificent zany" (Whipple, *World-Telegram*), "a magician in the happy realms of nonsense" (Brown, *Post*), "once more at the top of his form" (Watts, *Herald Tribune*), "a more perfect fool than ever" (Anderson, *Journal-American*), "a perfect fool out of a demented volume of Mother Goose" (Atkinson, *Times*).

Boys and Girls Together was Ed Wynn's last "legitimate" show on Broadway, but he did return in the summer of 1942 as the chief attraction of *Laugh, Town, Laugh,* one of the dozen or so wartime efforts to revive vaudeville. Though the show gave twelve performances a week at the top ticket price of $2.20, and was economically produced without new songs, elaborate scenery, or a chorus line, it was much the same Ed Wynn mixture as before. Ed made his first appearance emerging from between the halves of a huge hot-dog bun ("I just stepped out of a new roll"), and offered a lineup of talent that allowed him to cavort with singers (he accompanied Jane Froman on his piano-tricycle), trapeze artists, a ventriloquist, trained dogs, a Russian singing chorus, and two badminton experts. He also revived his duck-shooting specialty from *Boys and Girls Together*. Despite generally favorable reviews, *Laugh, Town, Laugh* ran only a month at the Alvin, and Ed took his final Broadway bow on July 25, 1942. (He attempted comebacks in two other similar vaudeville shows, *Big*

Time in 1943 and *Ed Wynn's Laugh Carnival* in 1948, but neither one reached New York.)

Wynn enjoyed some success as a television headliner in variety shows between 1949 and 1953, but after his contract was cancelled he was forced to spend two humiliating years job-hunting. His fortune, however, was reversed in 1956 when, at the age of sixty-nine, he played his first dramatic role in the film *The Great Man*. This was followed by an even more challenging assignment in the television play, *Requiem for a Heavyweight* (which was shown before *The Great Man* was released). Suddenly in demand as an actor, Ed Wynn was able to end his years productively in a variety of major roles on television and in such films as *Marjorie Morningstar, The Diary of Anne Frank,* and *Mary Poppins.* He died at the age of seventy-nine in Beverly Hills, on June 19, 1966.

"A comedian is not a man who says funny things," Ed once remarked. "A comedian is one who says things funny. For instance, a comedian doesn't open funny doors—a comedian opens doors in a funny way." Ed Wynn spent almost a lifetime opening doors in a funny way.

Stage Credits

This section contains credits for stage productions, both musical and non-musical, that included appearances by the performers discussed in this book. These credits cover Broadway, London, and the road, but do not cover summer, regional, or repertory theatres.

Fanny Brice

1910. FOLLIES OF 1910 (4th Edition)

Revue with sketches by Harry B. Smith; music by Irving Berlin, Will Marion Cook, Gus Edwards, etc.; lyrics by Irving Berlin, Joe Jordan, Harry B. Smith, etc. Produced by Florenz Ziegfeld; directed by Julian Mitchell.

Cast included: Bert Williams, Rosie Green, Lillian Lorraine, George Bickel, Julian Mitchell.

New York run: Jardin de Paris, June 20, 1910; 88 performances.

1911. ZIEGFELD FOLLIES (5th Edition)

Revue with sketches by George V. Hobart; music by Maurice Levi, Raymond Hubbell, Irving Berlin, etc.; lyrics by George V. Hobart, Irving Berlin. Produced by Florenz Ziegfeld; directed by Julian Mitchell; dances by Gus Sohlke, Jack Mason.

Cast included: Leon Errol, Dolly Sisters, Bessie McCoy, Bert Williams, Lillian Lorraine, George White.

New York run: Jardin de Paris, June 26, 1911; 80 performances.

1912. THE WHIRL OF SOCIETY

Musical comedy with book by Harrison Rhodes; music by Louis A. Hirsch; lyrics by Harold Atteridge. Produced by the Messrs. Shubert; directed by J.C. Huffman; dances by William Wilson.

Cast included: Gaby Deslys, Al Jolson, Ada Lewis, Oscar Schwarz (Shaw), Melville Ellis, Lawrence D'Orsay.

NOTE: Miss Brice and the above-mentioned cast members were in the touring company of the musical, which opened at the Belasco Theatre, Washington, Nov. 18, 1912. The original Broadway production opened at the Winter Garden, March 5, 1912, and ran for 136 performances.

1913. HONEYMOON EXPRESS

Musical comedy with book & lyrics by Joseph Herbert & Harold Atteridge; music by Jean Schwartz. Produced by the Messrs. Shubert; directed by Ned Wayburn.

Cast included: Gaby Deslys, Al Jolson, Harry Fox, Yancsi Dolly, Harland Dixon.

New York run: Winter Garden, Feb. 6, 1913; 156 performances.

NOTE: Miss Brice was succeeded by Ina Claire after two months.

1915. NOBODY HOME

Musical comedy with book by Guy Bolton; music by Jerome Kern; lyrics mostly by Schuyler Greene, Herbert Reynolds. Produced by F. Ray Comstock & Elisabeth Marbury; directed by J. H. Benrimo; dances by David Bennett.

Cast included: Lawrence Grossmith, Charles Judels, Helen Clarke.

New York run: Princess Theatre, April 20, 1915; 135 performances.

NOTE: Miss Brice succeeded Adele Rowland in Nov. 1915, and toured in the show.

1916. ZIEGFELD FOLLIES (10th Edition)

Revue with sketches by Gene Buck, George V. Hobart; music mostly by Louis A. Hirsch, David Stamper, Jerome Kern, Blanche Merrill; lyrics by Gene Buck, George V. Hobart, Blanche Merrill. Produced by Florenz Ziegfeld; directed by Ned Wayburn.

Cast included: W. C. Fields, Ann Pennington, Frances White, Bernard Granville, Justine Johnstone, Ina Claire, Bert Williams, Lilyan Tashman, Sam Hardy, Marion Davies, Carl Randall, Allyn King.

New York run: New Amsterdam Theatre, June 12, 1916; 112 performances.

1917. ZIEGFELD FOLLIES (11th Edition)

Revue with sketches by Gene Buck, George V. Hobart; music by Raymond Hubbell, David Stamper, Victor Herbert, Blanche Merrill; lyrics by Gene Buck, George V. Hobart, Blanche Merrill. Produced by Florenz Ziegfeld; directed by Ned Wayburn.

Cast included: W. C. Fields, Eddie Cantor, Will Rogers, Peggy Hopkins, Walter Catlett, Bert Williams, Fairbanks Twins, Lilyan Tashman, Carl Hyson & Dorothy Dickson, Allyn King.

New York run: New Amsterdam Theatre, June 12, 1917; 111 performances.

1918. WHY WORRY?

Play by Montague Glass & Jules Eckert Goodman. Produced by A. H. Woods; directed by George Marion.

Cast included: May Boley, George Sidney, Smith & Dale, Edwin Maxwell.
New York run: Harris Theatre, Aug. 23, 1918; 27 performances.

1918. ZIEGFELD NINE O'CLOCK REVUE

Cabaret revue with sketches by Gene Buck; music by David Stamper, Blanche Merrill; lyrics by Gene Buck, Blanche Merrill. Produced by Florenz Ziegfeld; directed by Ned Wayburn.

Cast included: Bert Williams, Lillian Lorraine, Van & Schenck, Johnny Dooley, Delyle Alda, Georgie Price.
New York run: New Amsterdam Roof, Oct. 18, 1918.

1919. ZIEGFELD MIDNIGHT FROLIC (8th Edition)

Cabaret revue with sketches & lyrics by Gene Buck; music by David Stamper. Produced by Florenz Ziegfeld; directed by Ned Wayburn.

Cast included: W. C. Fields, Frances White, Ted Lewis, Allyn King, Chic Sale, Savoy & Brennan.
New York run: New Amsterdam Roof, Oct. 2, 1919; 78 performances.

1920. ZIEGFELD GIRLS OF 1920

Cabaret revue with sketches by Gene Buck; music by David Stamper, James Hanley, Blanche Merrill; lyrics by Gene Buck, Ballard MacDonald, Blanche Merrill. Produced by Florenz Ziegfeld; directed by Ned Wayburn.

Cast included: W. C. Fields, Lillian Lorraine, Mary Hay, Allyn King, John Price Jones, Kathleen Martyn.
New York run: New Amsterdam Roof, March 8, 1920; 78 performances.

1920. ZIEGFELD MIDNIGHT FROLIC (9th Edition)

Cabaret revue with sketches by Gene Buck; music by David Stamper, Harry Carroll; lyrics by Gene Buck, Ballard MacDonald. Produced by Florenz Ziegfeld; directed by Ned Wayburn.

Cast included: W. C. Fields, Lillian Lorraine, Allyn King, John Price Jones, Kathleen Martyn, Carl Randall.
New York run: New Amsterdam Roof; March 1920.

1920. ZIEGFELD FOLLIES (14th Edition)

Revue with sketches by James Montgomery, George V. Hobart, W. C. Fields; music & lyrics mostly by Irving Berlin. Produced by Florenz Ziegfeld; directed by Edward Royce.

Cast included: W. C. Fields, Charles Winninger, Ray Dooley, Bernard Granville, John Steel, Van & Schenck, Carl Randall, Jack Donahue, Moran & Mack, Mary Eaton.

New York run: New Amsterdam Theatre, June 22, 1920; 123 performances.

1921. ZIEGFELD FOLLIES (15th Edition)

Revue with sketches by Willard Mack, Bud Spence, W. C. Fields; music by Rudolf Friml, David Stamper, Victor Herbert, Leo Edwards, Maurice Yvain; lyrics by Gene Buck, B. G. DeSylva, Channing Pollock, Ballard MacDonald. Produced by Florenz Ziegfeld; directed by Edward Royce, George Marion.

Cast included: W. C. Fields, Raymond Hitchcock, Ray Dooley, Mary Milburn, Van & Schenck, Florence O'Denishawn, Vera Michelena, Mary Eaton.

New York run: Globe Theatre, June 21, 1921; 119 performances.

1923. ZIEGFELD FOLLIES (18th Edition)

Revue with sketches by Gene Buck, Eddie Cantor; music by David Stamper, Victor Herbert, Leo Edwards, Harry Ruby; lyrics by Gene Buck, Blanche Merrill, Bert Kalmar. Produced by Florenz Ziegfeld; directed by Ned Wayburn.

Cast included: Bert & Betty Wheeler, Olga Steck, Lew Hearn, Ann Pennington, Harland Dixon, James J. Corbett, Paul Whiteman orch.

New York run: New Amsterdam Theatre, Oct. 20, 1923; 233 performances.

1924. MUSIC BOX REVUE

Revue with sketches by Bert Kalmar & Harry Ruby, Clark & McCullough, etc.; music & lyrics by Irving Berlin. Produced by Sam H. Harris; directed by John Murray Anderson; dances by Carl Randall, Mme. Serova.

Cast included: Bobby Clark & Paul McCullough; Oscar Shaw, Grace Moore, Carl Randall, Claire Luce, Brox Sisters.

New York run: Music Box, Dec. 1, 1924; 187 performances.

1926. FANNY

Play by Willard Mack & David Belasco. Produced & directed by David Belasco.

Cast included: Warren William, Spencer Charters, Jane Ellison.

New York run: Lyceum Theatre, Sept. 21, 1926; 63 performances.

1929. FIORETTA

Musical comedy with book by Earl Carroll & Charlton Andrews; music & lyrics mostly by G. Romilli (Romilly Johnson) & George Bagby. Produced & directed by Earl Carroll; dances by LeRoy Prinz.

Cast included: Leon Errol, Lionel Atwill, Dorothy Knapp, George Houston.

New York run: Earl Carroll Theatre, Feb. 5, 1929; 111 performances.

NOTE: Miss Brice was succeeded by Josephine Harmon on April 29, 1929.

1930. SWEET AND LOW

Revue with sketches by David Freedman; music by Will Irwin, Harry Warren, etc.; lyrics by Billy Rose, Mort Dixon, Charlotte Kent, Ira Gershwin, etc. Produced by Billy Rose; directed by Alexander Leftwich; dances by Danny Dare, Busby Berkeley.

Cast included: George Jessel, James Barton, Borrah Minevitch, Moss & Fontana, Hannah Williams, Arthur Treacher.
New York run: 46th St. Theatre, Nov. 17, 1930; 184 performances.

1931. BILLY ROSE'S CRAZY QUILT

Revue with sketches by David Freedman; music mostly by Harry Warren; lyrics mostly by Mort Dixon, Billy Rose. Produced & directed by Billy Rose; dances by Sammy Lee.

Cast included: Phil Baker, Ted Healy, Gomez & Winona, Lew Brice, Tamara.
New York run: 44th St. Theatre, May 19, 1931; 79 performances.

1934. ZIEGFELD FOLLIES

Revue with sketches by H. I. Phillips, Fred Allen, Harry Turgend, David Freedman; music mostly by Vernon Duke; lyrics mostly by E. Y. Harburg. Produced by Mrs. Florenz Ziegfeld & the Messrs. Shubert (uncredited); directed by Bobby Connolly, John Murray Anderson, Edward Clarke Lilley; dances by Robert Alton.

Cast included: Willie & Eugene Howard, Everett Marshall, Jane Froman, Vilma & Buddy Ebsen, Patricia Bowman, Cherry & June Preisser, Eve Arden, Brice Hutchins (Robert Cummings), Ina Ray (Hutton).
New York run: Winter Garden, Jan. 4, 1934; 182 performances.

1936. ZIEGFELD FOLLIES

Revue with sketches by David Freedman; music by Vernon Duke; lyrics by Ira Gershwin. Produced by Mrs. Florenz Ziegfeld & the Messrs. Shubert (uncredited); directed by John Murray Anderson, Edward Clarke Lilley; dances by Robert Alton, George Balanchine.

Cast included: Bob Hope, Josephine Baker, Gertrude Niesen, Hugh O'Connell, Harriet Hoctor, Cherry & June Preisser, Eve Arden, Judy Canova, John Hoysradt, Nicholas Brothers.
New York run: Winter Garden, Jan. 30, 1936; 115 performances.

1936. THE NEW ZIEGFELD FOLLIES OF 1936–1937

Cast included: Bobby Clark, Jane Pickens, Cherry & June Preisser, Gypsy Rose Lee, Cass Daley, Ruth Harrison & Alex Fisher.
New York run: Winter Garden, Sept. 14, 1936; 112 performances.

Bobby Clark

(All performances before 1936 were with Paul McCullough.)

1922. CHUCKLES OF 1922

Revue with sketches by William K. Wells, Clark & McCullough; music by Harry Archer; lyrics by P. D. Cook. Produced by Charles B. Cochran; dances by Seymour Felix.

Cast included: Emily Earle, Ruth Wheeler.
London run: New Oxford Theatre, June 19, 1922; 95 performances.

1922. MUSIC BOX REVUE

Revue with sketches by George V. Hobart, Walter Catlett, Paul Gerard Smith; music & lyrics by Irving Berlin. Produced by Sam H. Harris; directed by Hassard Short; dances by William Seabury.

Cast included: Charlotte Greenwood, William Gaxton, Fairbanks Twins, McCarthy Sisters, John Steel.
New York run: Music Box, Oct. 23, 1922; 330 performances.

1924. MUSIC BOX REVUE

Revue with sketches by Bert Kalmar & Harry Ruby, Clark & McCullough, etc.; music & lyrics by Irving Berlin. Produced by Sam H. Harris; directed by John Murray Anderson; dances by Carl Randall, Mme. Serova.

Cast included: Fannie Brice, Oscar Shaw, Grace Moore, Carl Randall, Claire Luce, Brox Sisters.
New York run: Music Box, Dec. 1, 1924; 187 performances.

1926. THE RAMBLERS

Musical comedy with book by Bert Kalmar, Harry Ruby, Guy Bolton; music by Harry Ruby; lyrics by Bert Kalmar. Produced by Philip Goodman; directed by John Harwood; dances by Sammy Lee.

Cust included: Jack Whiting, Marie Saxon, Bob Hope, Lew Parker.
New York run: Lyric Theatre, Sept. 20, 1926; 291 performances.

1930. STRIKE UP THE BAND

Musical comedy with book by Morrie Ryskind, George S. Kaufman; music by George Gershwin; lyrics by Ira Gershwin. Produced by Edgar Selwyn; directed by Alexander Leftwich; dances by George Hale.

Cast included: Blanche Ring, Doris Carson, Dudley Clements, Red Nichols orch.

(incl. Benny Goodman, Gene Krupa, Glenn Miller, Jimmy Dorsey, Jack Teagarden).
New York run: Times Square Theatre, Jan. 14, 1930; 191 performances.

1931. COCHRAN'S 1931 REVUE

Revue with sketches by Noël Coward, Ronald Jeans, etc.; music & lyrics by Noël Coward, etc. Produced by Charles B. Cochran; directed by Frank Collins; dances by Buddy Bradley.

Cast included: Ada May, John Mills, Buddy Bradley, Queenie Leonard.
London run: Pavilion Theatre, March 19, 1931; 27 performances.

1931. HERE GOES THE BRIDE

Musical comedy with book by Peter Arno; music by John W. Green; lyrics by Edward Heyman. Produced by Peter Arno; directed by Edward Clarke Lilley; dances by Russell Markert.

Cast included: Grace Brinkley, Frances Langford, Paul Frawley, Eric Blore, Victoria (Vicki) Cummings.
New York run: 46th St. Theatre, Nov. 3, 1931; 7 performances.

1932. WALK A LITTLE FASTER

Revue with sketches by S. J. Perelman, Robert MacGunigle; music by Vernon Duke; lyrics by E. Y. Harburg. Produced by Courtney Burr; directed by Monty Woolley; dances by Albertina Rasch.

Cast included: Beatrice Lillie, John Hundley, Evelyn Hoey, Donald Burr, Dorothy McNulty (Penny Singleton), Dave & Dorothy Fitzgibbon.
New York run: St. James Theatre, Dec. 7, 1932; 119 performances.

1934. THUMBS UP!

Revue with sketches by H. I. Phillips, Alan Baxter, Ronald Jeans, etc.; music mostly by James F. Hanley, Henry Sullivan; lyrics mostly by Ballard MacDonald, Earle Crooker. Produced by Eddie Dowling; directed by John Murray Anderson, Edward Clarke Lilley; dances by Robert Alton.

Cast included: Eddie Dowling, Hal LeRoy, J. Harold Murray, Eddie Garr, Ray Dooley, Pickens Sisters, Paul Draper, Rose King, Sheila Barrett, Jack Cole, Eunice Healey, Barnett Parker, John Fearnley.
New York run: St. James Theatre, Dec. 27, 1934; 156 performances.

1936. THE NEW ZIEGFELD FOLLIES OF 1936–1937

Revue with sketches by David Freedman; music by Vernon Duke; lyrics by Ira Gershwin. Produced by Mrs. Florenz Ziegfeld & the Messrs. Shubert (uncredited); directed by John Murray Anderson, Edward Clarke Lilley; dances by Robert Alton, George Balanchine.

Cast included: Fannie Brice, Jane Pickens, Cherry & June Preisser, Gypsy Rose Lee, Cass Daley, Alex Harrison & Ruth Fisher.
New York run: Winter Garden, Sept. 14, 1936; 112 performances.
NOTE: Mr. Clark was only in this fall edition of the 1936 *Follies*, which had opened at the Winter Garden on Jan. 30, 1936, and ran for 115 performances.

1939. THE STREETS OF PARIS

Revue with sketches by Frank Eyton, Tom McKnight, Edward Duryea Dowling, Charles Sherman, etc.; music by Jimmy McHugh; lyrics by Al Dubin. Produced by the Messrs. Shubert with Olsen and Johnson; directed by Edward Duryea Dowling; dances by Robert Alton.

Cast included: Luella Gear, Bud Abbott & Lou Costello, Jean Sablon, Carmen Miranda, Gower Champion & Jeanne Tyler, John (Jack) McCauley, Hugh Martin, Della Lind.
New York run: Broadhurst Theatre, June 19, 1939; 274 performances.

1940. LOVE FOR LOVE

Play by William Congreve. Produced by The Players; directed by Robert Edmond Jones.

Cast included: Cornelia Otis Skinner, Peggy Wood, Leo G. Carroll, Dudley Digges, Violet Heming, Dorothy Gish, Barry Jones, Romney Brent.
New York run: Hudson Theatre, June 3, 1940; 8 performances.

1941. ALL MEN ARE ALIKE

Play by Vernon Sylvaine. Produced by Lee Ephraim; directed by Harry Wagstaff Gribble.

Cast included: Reginald Denny, Cora Witherspoon, Lillian Bond.
New York run: Hudson Theatre, Oct. 6, 1941; 32 performances.

1942. THE RIVALS

Play by Richard Brinsley Sheridan. Produced by The Theatre Guild; directed by Eva Le Gallienne.

Cast included: Mary Boland, Walter Hampden, Helen Ford, Haila Stoddard, Philip Bourneuf.
New York run: Shubert Theatre, Jan. 14, 1942; 54 performances.

1942. STAR AND GARTER

Revue with sketches by H. I. Phillips, Bobby Clark, etc.; music by Jerry Seelen Irving Berlin, Harold Arlen, Jimmy McHugh, etc.; lyrics by Lester Lee, Irving Berlin, Johnny Mercer, Al Dubin, etc. Produced by Michael Todd; directed by Hassard Short; dances by Al White Jr.

Cast included: Gypsy Rose Lee, Georgia Sothern, Prof. Lamberti, Pat Harrington, Carrie Finnell.
New York run: Music Box, June 24, 1942; 609 performances.

1944. MEXICAN HAYRIDE

Musical comedy with book by Herbert & Dorothy Fields; music & lyrics by Cole Porter. Produced by Michael Todd; directed by Hassard Short, John Kennedy; dances by Paul Haakon.

Cast included: June Havoc, George Givot, Wilbur Evans, Luba Malina, Corinna Mura, Paul Haakon, Edith Meiser, Bill Callahan.
New York run: Winter Garden, Jan. 28, 1944; 481 performances.

1946. THE WOULD-BE GENTLEMAN

Play by Molière, adapted by Bobby Clark & William Roos (uncredited). Produced by Michael Todd; directed by John Kennedy.

Cast included: Edith King, Gene Barry, June Knight, Eleanore Whitney, Ruth Harrison & Alex Fisher.
New York run: Booth Theatre, Jan. 9, 1946; 77 performances.

1947. SWEETHEARTS

Musical comedy with book by Fred DeGresac & Harry B. Smith, revised by John Cecil Holm; music by Victor Herbert; lyrics by Robert B. Smith. Produced by Paula Stone & Michael Sloane; directed by John Kennedy; dances by Theodore Adolphus, Catherine Littlefield.

Cast included: Marjorie Gateson, June Knight, Mark Dawson, Gloria Story, Cornell MacNeil.
New York run: Shubert Theatre, Jan. 21, 1947; 288 performances.

1948. AS THE GIRLS GO

Musical comedy with book by William Roos; music by Jimmy McHugh; lyrics by Harold Adamson. Produced by Michael Todd; directed by Howard Bay; dances by Hermes Pan.

Cast included: Irene Rich, Betty Jane Watson, Kathryn Lee, Hobart Cavanaugh, Bill Callahan, Jo Sullivan, Abbe Marshall (Abbe Lane).
New York run: Winter Garden, Nov. 13, 1948; 420 performances.

1952. JOLLYANNA

Musical comedy with book by E. Y. Harburg & Fred Saidy; music by Sammy Fain; lyrics by E. Y. Harburg. Produced by the San Francisco & Los Angeles Civic Light Opera Assn., Edwin Lester, director; directed by Jack Donohue; dances by Ruthanna Boris.

Cast included: Mitzi Gaynor, John Beal, Biff McGuire, Marthe Erroll, Beverly Tyler, Bil Baird Marionettes.

New York run: none.

NOTE: Musical opened at the Curran Theatre, San Francisco, Sept. 11, 1952, and closed at the Philharmonic Auditorium, Los Angeles, Oct. 4, 1952.

1956. DAMN YANKEES

Musical comedy with book by George Abbott & Douglas Wallop; music & lyrics by Richard Adler & Jerry Ross. Produced by Frederick Brisson, Richard Griffith & Harold Prince; directed by George Abbott; dances by Bob Fosse.

Cast included: Sherry O'Neil, Allen Case, Sid Stone, Rosemary Kuhlmann.

NOTE: Mr. Clark and the above-mentioned cast members were in the touring company of the musical, which opened at the Shubert Theatre, New Haven, Jan. 29, 1956, and closed at the Royal Alexandra Theatre, Toronto, May 18, 1957. The original Broadway production opened at the 46th St. Theatre, May 5, 1955, and ran for 1,022 performances.

Joe Cook

1918. HALF PAST EIGHT

Revue with sketches uncredited; music mostly by George Gershwin; lyrics mostly by Edward B. Perkins. Produced by Edward B. Perkins.

Cast included: Sybil Vane, Roy Stever & Mildred Lovejoy, Ruby Loraine.

New York run: None.

NOTE: Show opened at the Empire Theatre, Syracuse, Dec. 9, 1918, and played 6 performances.

1919. HITCHY-KOO, 1919

Revue with sketches by George V. Hobart; music & lyrics by Cole Porter. Produced by Raymond Hitchcock; directed by Julian Alfred.

Cast included: Raymond Hitchcock, Florence O'Denishawn, Billy Holbrook, Lillian Kemble Cooper.

New York run: Liberty Theatre, Oct. 6, 1919; 56 performances.

1923. VANITIES OF 1923 (1st Edition)

Reveue with sketches uncredited; music & lyrics by Earl Carroll. Produced & directed by Earl Carroll; dances by Sammy Lee, F. Renoff.

Cast included: Peggy Hopkins Joyce, J. Frank Leslie, Bernard Granville, Dorothy Knapp, Russell Markert.

New York run: Earl Carroll Theatre, July 5, 1923; 204 performances.

1924. EARL CARROLL VANITIES (2nd Edition)

Revue with sketches uncredited; music & lyrics by Earl Carroll. Produced & directed by Earl Carroll; dances by Sammy Lee.

Cast included: Sophie Tucker, Desirée Tabor, Dare & Wahl.
New York run: Music Box, Sept. 10, 1924; 133 performances.

1925. HOW'S THE KING?

Musical comedy with book by Marc Connelly; music by Jay Gorney; lyrics by Owen Murphy. Produced by Earl Carroll; directed by Robert Milton; dances by Allan K. Foster.

Cast included: John Price Jones, J. M. Kerrigan.
New York run: None.
NOTE: Musical opened at the Chestnut St. Opera House, Philadelphia, Sept. 12, 1925, where it played one week.

1925. EARL CARROLL VANITIES (4th Edition)

Revue with sketches mostly by William A. Grew; music & lyrics mostly by Clarence Gaskill. Produced & directed by Earl Carroll; dances by David Bennett.

Cast included: Frank Tinney, Julius Tannen, Ted Healy, Dave Chasen, Jessica Dragonette, Red Nichols orch.
New York run: Earl Carroll Theatre, Dec. 28, 1925; 230 performances.
NOTE: The 4th Edition of the *Vanities* was a continuation of the 3rd Edition, which had opened July 6, 1925. Total run of both editions was 429 performances.

1928. RAIN OR SHINE

Musical comedy with book by James Gleason & Maurice Marks, music by Milton Ager; lyrics by Jack Yellen, Owen Murphy. Produced by A. L. Jones & Morris Green; directed by Alexander Leftwich; dances by Russell Markert, Tom Nip.

Cast included: Tom Howard, Nancy Welford, Dave Chasen, Warren Hull.
New York run: George M. Cohan Theatre, Feb. 9, 1928; 360 performances.

1930. FINE AND DANDY

Musical comedy with book by Donald Ogden Stewart; music by Kay Swift; lyrics by Paul James. Produced by Morris Green & Lewis Gensler; directed by Morris Green, Frank McCoy; dances by Dave Gould, Tom Nip.

Cast included: Eleanor Powell, Nell O'Day, Dave Chasen, Alice Boulden, Joe Wagstaff.
New York run: Erlanger's Theatre, Sept. 23, 1930; 255 performances.

1932. FANFARE

Revue with sketches by Dion Titheradge, Beverly Nichols; music by Henry Sullivan; lyrics by Desmond Carter. Produced & directed by John Murray Anderson.

Cast included: June, Violet Loraine, Joyce Barbour, Dave Chasen, Condos Brothers, Bernard Clifton.
London run: Prince Edward Theatre, June 23, 1932.

1933. HOLD YOUR HORSES

Musical comedy with book by Russel Crouse & Corey Ford; music by Robert Russell Bennett; lyrics by Robert A. Simon & Owen Murphy. Produced by the Messrs. Shubert (uncredited); directed by R. H. Burnside, John Shubert; dances by Robert Alton, Harriet Hoctor.

Cast included: Ona Munson, Tom Patricola, Dave Chasen, Rex Weber, Harriet Hoctor, Stanley Smith, Jack Powell.
New York run: Winter Garden, Sept. 25, 1933; 88 performances.

1939. OFF TO BUFFALO

Play by Max Liebman & Allen Boretz. Produced by Albert Lewis; directed by Melville Burke.

Cast included: Hume Cronyn, Matt Briggs, Otto Hulett.
New York run: Ethel Barrymore Theatre, Feb. 21, 1939; 7 performances.

1940. IT HAPPENS ON ICE

Ice revue with music by Vernon Duke, Fred Ahlert, Peter DeRose; lyrics by Al Stillman, Mitchell Parish. Produced by Sonja Henie & Arthur Wirtz; directed by Leon Leonidoff; dances by Catherine Littlefield.

Cast included: Joan Edwards, Felix Knight, Jack Kilty.
New York run: Center Theatre, Oct. 10, 1940; 180 performances.
NOTE: Show reopened without Mr. Cook on April 4, 1941. Total run was 662 performances.

Jimmy Durante

1929. SHOW GIRL

Musical comedy with book by William Anthony McGuire; music by George Gershwin; lyrics by Ira Gershwin & Gus Kahn. Produced by Florenz Ziegfeld; directed by William Anthony McGuire; dances by Bobby Connolly, Albertina Rasch.

Cast included: Ruby Keeler, Lou Clayton & Eddie Jackson, Eddie Foy Jr., Harriet Hoctor, Frank McHugh, Doris Carson, Barbara Newberry, Duke Ellington orch.
New York run: Ziegfeld Theatre, July 2, 1929; 111 performances.

1930. THE NEW YORKERS

Musical comedy with book by Herbert Fields; music & lyrics by Cole Porter. Produced by E. Ray Goetz; directed by Monty Woolley; dances by George Hale.

Cast included: Charles King, Frances Williams, Hope Williams, Ann Pennington, Richard Carle, Marie Cahill, Lou Clayton & Eddie Jackson, Fred Waring Pennsylvanians, Kathryn Crawford.
New York run: Broadway Theatre, Dec. 8, 1930; 168 performances.

1933. STRIKE ME PINK

Revue with sketches by Lew Brown & Ray Henderson; music by Ray Henderson; lyrics by Lew Brown. Produced & directed by Lew Brown & Ray Henderson; dances by Seymour Felix.

Cast included: Lupe Velez, Hope Williams, Hal LeRoy, Roy Atwell, Eddie Garr, Ruth Harrison & Alex Fisher, Gracie Barrie, Johnny Downs, George Dewey Washington.
New York run: Majestic Theatre, March 4, 1933; 105 performances.

1935. JUMBO

Musical comedy with book by Ben Hecht & Charles MacArthur; music by Richard Rodgers; lyrics by Lorenz Hart. Produced by Billy Rose; directed by John Murray Anderson, George Abbott; dances by Allan K. Foster.

Cast included: Paul Whiteman orch., Donald Novis, Gloria Grafton, A. P. Kaye, A. Robins, Poodles Hanneford.
New York run: Hippodrome, Nov. 16, 1935; 233 performances.

1936. RED, HOT AND BLUE!

Musical comedy with book by Howard Lindsay & Russel Crouse; music & lyrics by Cole Porter. Produced by Vinton Freedley; directed by Howard Lindsay; dances by George Hale.

Cast included: Ethel Merman, Bob Hope, Paul & Grace Hartman, Vivian Vance, Polly Walters, Lew Parker.
New York run: Alvin Theatre, Oct. 29, 1936; 183 performances.

1939. STARS IN YOUR EYES

Musical comedy with book by J. P. McEvoy; music by Arthur Schwartz; lyrics by Dorothy Fields. Produced by Dwight Deere Wiman; directed by Joshua Logan; dances by Carl Randall.

Cast included: Ethel Merman, Tamara Toumanova, Richard Carlson, Mildred Natwick, Dan Dailey, Jr., Mary Wickes, Nora Kaye, Jerome Robbins, Alicia Alonso, Paul Godkin.

New York run: Majestic Theatre, Feb. 9, 1939; 127 performances.

1940. KEEP OFF THE GRASS

Revue with sketches by Parke Levy, Alan Lipscott, S. J. Kaufman, etc.; music by Jimmy McHugh; lyrics by Al Dubin, Howard Dietz. Produced by the Messrs. Shubert; directed by Fred DeCordova, Edward Duryea Dowling; dances by George Balanchine.

Cast included: Ray Bolger, Jane Froman, Ilka Chase, Betty Bruce, Jackie Gleason, José Limon, Larry Adler, Virginia O'Brien.

New York run: Broadhurst Theatre, May 23, 1940; 44 performances.

W. C. Fields

1905. THE HAM TREE

Musical comedy with book by George V. Hobart; music by Jean Schwartz; lyrics by William Jerome. Produced by Klaw & Erlanger; directed by Herbert Gresham, Ned Wayburn.

Cast included: James McIntyre & T. K. Heath, David Torrence, Jobyna Howland.

New York run: New York Theatre, Aug. 28, 1905; 90 performances.

1914. WATCH YOUR STEP

Musical comedy with book by Harry B. Smith, music & lyrics by Irving Berlin. Produced by Charles B. Dillingham; directed by R. H. Burnside.

Cast included: Vernon & Irene Castle, Frank Tinney, Charles King, Elizabeth Brice, Elizabeth Murray, Harry Kelly, Justine Johnstone.

NOTE: Mr. Fields appeared in this musical for one performance only, Nov. 25, 1914, during its tryout at the Empire Theatre, Syracuse. The production then opened on Broadway at the New Amsterdam Theatre, Dec. 8, 1914, and ran for 175 performances.

1915. ZIEGFELD FOLLIES (9th Edition)

Revue with sketches & lyrics by Channing Pollock, Rennold Wolf, Gene Buck; music by Louis A. Hirsch, David Stamper. Produced by Florenz Ziegfeld; directed by Julian Mitchell, Leon Errol.

Cast included: Ed Wynn, Ina Claire, Ann Pennington, Justine Johnstone, Mae Murray, Bernard Granville, Bert Williams, George White, Leon Errol, Carl Randall.

New York run: New Amsterdam Theatre, June 21, 1915; 104 performances.

1916. ZIEGFELD FOLLIES (10th Edition)

Revue with sketches & lyrics by Gene Buck, George V. Hobart; music mostly by Louis A. Hirsch, Jerome Kern, David Stamper. Produced by Florenz Ziegfeld; directed by Ned Wayburn.

Cast included: Fannie Brice, Ann Pennington, Frances White, Bernard Granville, Justine Johnstone, Ina Claire, Bert Williams, Lilyan Tashman, Sam Hardy, Marion Davies, Carl Randall, Allyn King.
New York run: New Amsterdam Theatre, June 12, 1916; 112 performances.

1917. ZIEGFELD FOLLIES (11th Edition)

Revue with sketches & lyrics by Gene Buck, George V. Hobart; music by Raymond Hubbell, David Stamper, Victor Herbert. Produced by Florenz Ziegfeld; directed by Ned Wayburn.

Cast included: Fannie Brice, Eddie Cantor, Will Rogers, Peggy Hopkins, Walter Catlett, Bert Williams, Fairbanks Twins, Lilyan Tashman, Carl Hyson & Dorothy Dickson, Allyn King.
New York run: New Amsterdam Theatre, June 12, 1917; 111 performances.

1918. ZIEGFELD FOLLIES (12th Edition)

Revue with sketches by Rennold Wolf, Gene Buck, W.C. Fields; music by Louis A. Hirsch, David Stamper, Irving Berlin. Produced by Florenz Ziegfeld; directed by Ned Wayburn.

Cast included: Eddie Cantor, Will Rogers, Marilyn Miller, Ann Pennington, Lillian Lorraine, Harry Kelly, Fairbanks Twins, Allyn King.
New York run: New Amsterdam Theatre, June 18, 1918; 151 performances.

1919. ZIEGFELD MIDNIGHT FROLIC (8th Edition)

Cabaret revue with sketches & lyrics by Gene Buck; music by David Stamper. Produced by Florenz Ziegfeld; directed by Ned Wayburn.

Cast included: Fannie Brice, Frances White, Ted Lewis, Allyn King, Chic Sale, Savoy & Brennan.
New York run: New Amsterdam Roof, Oct. 2, 1919; 78 performances.

1920. ZIEGFELD GIRLS OF 1920

Cabaret revue with sketches & lyrics by Gene Buck; music by David Stamper. Produced by Florenz Ziegfeld; directed by Ned Wayburn.

Cast included: Fannie Brice, Lillian Lorraine, Mary Hay, Allyn King, John Price Jones, Kathleen Martyn.
New York run: New Amsterdam Roof, March 8, 1920; 78 performances.

1920. ZIEGFELD MIDNIGHT FROLIC (9th Edition)

Cabaret revue with sketches by Gene Buck; music by David Stamper, Harry Carroll; lyrics by Gene Buck, Ballard MacDonald. Produced by Florenz Ziegfeld; directed by Ned Wayburn.

Cast included: Fannie Brice, Lillian Lorraine, Allyn King, John Price Jones, Kathleen Martyn, Carl Randall.
New York run: New Amsterdam Roof, March 1920.

1920. ZIEGFELD FOLLIES (14th Edition)

Revue with sketches by James Montgomery, George V. Hobart, W. C. Fields; music & lyrics mostly by Irving Berlin. Produced by Florenz Ziegfeld; directed by Edward Royce.

Cast included: Fannie Brice, Charles Winninger, Ray Dooley, Bernard Granville, John Steel, Van & Schenck, Carl Randall, Jack Donahue, Moran & Mack, Mary Eaton.
New York run: New Amsterdam Theatre, June 22, 1920; 123 performances.

1921. ZIEGFELD FOLLIES (15th Edition)

Revue with sketches by Willard Mack, Bud Spence, W. C. Fields; music mostly by Rudolf Friml, David Stamper, Victor Herbert; lyrics mostly by Gene Buck, B. G. DeSylva, Channing Pollock. Produced by Florenz Ziegfeld; directed by Edward Royce, George Marion.

Cast included: Fannie Brice, Raymond Hitchcock, Ray Dooley, Mary Milburn, Van & Schenck, Florence O'Denishawn, Vera Michelena, Mary Eaton.
New York run: Globe Theatre, June 21, 1921; 119 performances.

1922. GEORGE WHITE'S SCANDALS (4th Edition)

Revue with sketches by W. C. Fields, Andy Rice, George White; music by George Gershwin; lyrics by B. G. DeSylva & E. Ray Goetz. Produced & directed by George White.

Cast included: Winnie Lightner, Jack McGowan, George White, Paul Whiteman orch.
New York run: Globe Theatre, Aug. 28, 1922; 88 performances.

1923. POPPY

Musical comedy with book by Dorothy Donnelly with Howard Dietz & W. C. Fields (uncredited); lyrics by Dorothy Donnelly; music by Stephen Jones, Arthur Samuels. Produced by Philip Goodman; directed by Dorothy Donnelly; dances by Julian Alfred.

Cast included: Madge Kennedy, Luella Gear, Robert Woolsey.
New York run: Apollo Theatre, Sept. 3, 1923; 346 performances.

1925. THE COMIC SUPPLEMENT

Revue with sketches by J. P. McEvoy & W. C. Fields; lyrics by J. P. McEvoy; music by Con Conrad, Henry Souvaine. Produced by Florenz Ziegfeld; directed by Augustin Duncan; dances by Julian Mitchell.

Cast included: Ray Dooley, Betty Compton, Clarence Nordstrom.
New York run: None.
NOTE: Revue opened at the Shubert Theatre, Newark, Jan. 26, 1925, where it played three weeks, then played Washington.

1925. ZIEGFELD FOLLIES (19th & 20th Editions)

Revue with sketches by J. P. McEvoy, W. C. Fields, Will Rogers; music by Raymond Hubbell, David Stamper, Werner Janssen; lyrics by Gene Buck. Produced by Florenz Ziegfeld; dieected by Julian Mitchell.

Cast included: Will Rogers, Vivienne Segal, Clarence Nordstrom, Ray Dooley, Bertha Belmore, Tom Lewis, George Olsen orch.
New York run: New Amsterdam Theatre, March 10, 1925.
NOTE: This was the "Spring Edition" of a revue that had opened Sept. 23, 1924, as the 19th Edition. By July 6, 1925, so many changes had been made that the revue was then called the 20th Edition. Total run of both editions was 510 performances.

1928. EARL CARROLL VANITIES (7th Edition)

Revue with sketches by W. C. Fields, Paul Gerard Smith; music & lyrics mostly by Grace Henry & Morris Hamilton. Produced & directed by Earl Carroll; dances by Busby Berkeley.

Cast included: Ray Dooley, Joe Frisco, Gordon Dooley, Lillian Roth, Dorothy Knapp, Vincent Lopez orch.
New York run: Earl Carroll Theatre, Aug. 6, 1928; 203 performances.

1930. BALLYHOO

Musical comedy with book & lyrics by Harry Ruskin & Leighton K. Brill; music by Louis Alter. Produced by Arthur Hammerstein; directed by Reginald Hammerstein; dances by Earl Lindsey.

Cast included: Grace Hayes, Janet Reade, Don Tomkins, Chaz Chase, Ted Black orch.
New York run: Hammerstein's Theatre, Dec. 22, 1930; 68 performance.

Willie Howard

1901. THE LITTLE DUCHESS

Musical comedy with book & lyrics by Harry B. Smith; music by Reginald DeKoven. Produced by Florenz Ziegfeld; directed by George Marion.

Cast included: Anna Held, Joseph Herbert, George Marion, Bessie Wynn, Joe Welch.
NOTE: Mr. Howard appeared in this musical for one performance only, during its tryout in Washington. The production then opened on Broadway at the Casino Theatre, Oct. 14, 1901, and ran for 136 performances.

1912. THE PASSING SHOW OF 1912

Revue with sketches & lyrics by Harold Atteridge; music by Louis A. Hirsch. Produced by the Messrs. Shubert; directed by Ned Wayburn.

Cast included: Eugene Howard, Harry Fox, Anna Wheaton, Ernest Hare, Charlotte Greenwood, Trixie Friganza, Oscar Shaw, Jobyna Howland.
New York run: Winter Garden, July 22, 1912; 136 performances.

1914. THE WHIRL OF THE WORLD

Revue with sketches & lyrics by Harold Atteridge; music by Sigmund Romberg. Produced by the Messrs. Shubert; directed by William J. Wilson; dances by Jack Mason.

Cast included: Eugene Howard, Bernard Granville, Walter C. Kelly, Roszika Dolly, Lydia Kyasht, Lillian Lorraine, Ralph Herz.
New York run: Winter Garden, Jan. 10, 1914; 161 performances.

1915. THE PASSING SHOW OF 1915

Revue with sketches & lyrics by Harold Atteridge; music by Leo Edwards, W. F. Peters, J. Leubrie Hill. Produced by the Messrs. Shubert; directed by J. C. Huffman; dances by Jack Mason, Theodor Kosloff.

Cast included: Eugene Howard, Marilynn Miller, John Charles Thomas, Sam Hearn, Ernest Hare, Theodor Kosloff.
New York run: Winter Garden, May 29, 1915; 145 performances.

1916. THE SHOW OF WONDERS

Revue with sketches & lyrics by Harold Atteridge; music by Sigmund Romberg, Otto Motzan, Herman Timberg. Produced by the Messrs. Shubert; directed by J. C. Huffman; dances by Allan K. Foster.

Cast included: Eugene Howard, McIntyre & Heath, Ernest Hare, Walter C. Kelly, Marilyn Miller, Lou Clayton & Sam White.
New York run: Winter Garden, Oct. 26, 1916; 209 performances.

1918. THE PASSING SHOW OF 1918

Revue with sketches & lyrics Harold Atteridge; music by Sigmund Romberg, Jean Schwartz. Produced by the Messrs. Shubert; directed by J. C. Huffman; dances by Jack Mason

Cast included: Eugene Howard, Fred & Adele Astaire, Charles Ruggles, Frank Fay, Lou Clayton & Sam White, Nita Naldi, George Hassell, Violet Englefield.
New York run: Winter Garden, July 25, 1918; 124 performances.

1920. THE PASSING SHOW OF 1921

Revue with sketches & lyrics by Harold Atteridge; music by Jean Schwartz. Produced by the Messrs. Shubert; directed by J. C. Huffman; dances by Max Scheck.

Cast included: Eugene Howard, Marie Dressler, Harry Watson, J. Harold Murray, Sammy White, Perry Askam, Violet Englefield, Ina Hayward.

New York run: Winter Garden, Dec. 29, 1920; 200 performances.

1922. THE PASSING SHOW OF 1922

Revue with sketches & lyrics by Harold Atteridge; music by Alfred Goodman. Produced by the Messrs. Shubert; directed by J. C Huffman; dances by Allan K. Foster.

Cast included: Eugene Howard, Fred Allen, Arthur Margetson, Ethel Shutta, George Hassell.

New York run: Winter Garden, Sept, 22, 1922; 95 performances.

1925. SKY HIGH

Musical comedy with book & lyrics by Harold Atteridge & Harry Graham; music by Robert Stolz, Alfred Goodman. Produced by the Messrs. Shubert; directed by Fred G. Latham, Alexander Leftwich; dances by Seymour Felix.

Cast included: James R. Liddy, Joyce Barbour, Florenz Ames, Vanessi.

New York run: Shubert Theatre, March 2, 1925; 217 performances.

1926. GEORGE WHITE'S SCANDALS (8th Edition)

Revue with sketches by George White, William K. Wells; music by Ray Henderson; lyrics by B. G. DeSylva & Lew Brown. Produced & directed by George White.

Cast included: Eugene Howard, Frances Williams, Tom Patricola, Harry Richman, Buster West, Ann Pennington, McCarthy Sisters, Fairbanks Twins, Portland Hoffa.

New York run: Apollo Theatre, June 14, 1926; 432 performances.

1928. GEORGE WHITE'S SCANDALS (9th Edition)

Revue with sketches by George White, William K. Wells; music by Ray Henderson; lyrics by B. G. DeSylva & Lew Brown. Produced & directed by George White; dances by George White, Russell Markert.

Cast included: Eugene Howard, Ann Pennington, Harry Richman, Tom Patricola, Frances Williams, Arnold Johnson orch. (with Harold Arlen).

New York run: Apollo Theatre, July 2, 1928; 240 performances.

1929. GEORGE WHITE'S SCANDALS (10th Edition)

Revue with sketches by George White, William K. Wells; music by Cliff Friend; lyrics by George White. Produced & directed by George White; dances by George White, Merriel Abbott.

Cast included: Eugene Howard, Frances Williams, Mitchell & Durant, Evelyn Wilson, George White.

New York run: Apollo Theatre, Sept. 23, 1929; 161 performances.

1930. GIRL CRAZY

Musical comedy with book by Guy Bolton & John McGowan; music by George Gershwin; lyrics by Ira Gershwin. Produced by Alex A. Aarons & Vinton Freedley; directed by Alexander Leftwich; dances by George Hale.

Cast included: Allen Kearns, Ginger Rogers, William Kent, Ethel Merman, Antonio & Renée DeMarco, Lew Parker, Eunice Healey, The Foursome, Red Nichols orch. (incl. Benny Goodman, Gene Krupa, Glenn Miller, Jimmy Dorsey, Jack Teagarden).

New York run: Alvin Theatre, Oct. 14, 1930; 272 performances.

1931. GEORGE WHITE'S SCANDALS (11th Edition)

Revue with sketches by George White, Lew Brown; music by Ray Henderson; lyrics by Lew Brown. Produced & directed by George White.

Cast included: Eugene Howard, Rudy Vallee, Ethel Merman, Everett Marshall, Ray Bolger, Ethel Barrymore Colt, Loomis Sisters, Alice Faye.

New York run: Apollo Theatre, Sept. 14, 1931; 202 performances.

1932. BALLYHOO OF 1932

Revue with sketches by Norman Anthony, Sig Herzig; music by Lewis Gensler; lyrics by E. Y. Harburg. Produced & directed by Norman Anthony, Lewis Gensler, Bobby Connolly, Russell Patterson; dances by Bobby Connolly.

Cast included: Eugene Howard, Jeanne Aubert, Lulu McConnell, Bob Hope, Paul & Grace Hartman, Nina Mae McKinney.

New York run: 44th St. Theatre, Sept. 6, 1932; 95 performances.

1933. GEORGE WHITE'S MUSIC HALL VARIETIES

Revue with sketches by William K. Wells & George White; music by Harold Arlen, Sam. H. Stept, etc.; lyrics by Irving Caesar, Herb Magidson, etc. Produced & directed by George White; dances by Russell Markert.

Cast included: Eugene Howard, Bert Lahr, Harry Richman, Tom Patricola, Eleanor Powell, Loomis Sisters, Betty Kean.

New York run: Casino Theatre, Nov. 22, 1932; 72 performances.

NOTE: Willie & Eugene Howard joined the cast Jan. 2, 1933.

1934. ZIEGFELD FOLLIES

Revue with sketches by H.I. Phillips, Fred Allen, Harry Turgend, David Freedman; music mostly by Vernon Duke; lyrics mostly by E.Y. Harburg. Produced by Mrs. Florenz Ziegfeld & the Messrs. Shubert (uncredited); directed by Bobby Connolly, John Murray Anderson, Edward Clark Lilley; dances by Robert Alton.

Cast included: Eugene Howard, Fannie Brice, Everett Marshall, Jane Froman, Vilma & Buddy Ebsen, Patricia Bowman, Cherry & June Preisser, Eve Arden, Brice Hutchins (Robert Cummings), Ina Ray (Hutton).
New York run: Winter Garden, Jan. 4, 1934; 182 performances.

1935. GEORGE WHITE'S SCANDALS (12th Edition)

Revue with sketches by William K. Wells, George White, Howard Shiebler; music by Ray Henderson; lyrics by Jack Yellen. Produced & directed by George White; dances by Russell Markert.

Cast included: Eugene Howard, Bert Lahr, Rudy Vallee, Gracie Barrie, Cliff Edwards, Jane Cooper, Sam, Ted & Ray, Hal Forde.
New York run: New Amsterdam Theatre, Dec. 25, 1935; 110 performances.

1937. THE SHOW IS ON

Revue with sketches by David Freedman, Moss Hart; music mostly by Vernon Duke; lyrics mostly by Ted Fetter. Produced by the Messrs. Shubert; directed by Vincente Minnelli, Edward Clark Lilley; dances by Robert Alton.

Cast included: Eugene Howard, Rose King, Chic York, Roy Cropper, Jack McCauley.
New York run: Winter Garden, Sept. 18, 1937; 17 performances.
NOTE: This was return engagement (prior to tour) of revue that had opened at the Winter Garden on Dec. 25, 1936, and ran for 236 performances.

1939. GEORGE WHITE'S SCANDALS (13th Edition)

Revue with sketches by Matt Brooks, Eddie Davis, George White; music by Sammy Fain; lyrics by Jack Yellen. Produced & directed by George White; dances by George White.

Cast included: Eugene Howard, The Three Stooges, Ben Blue, Ella Logan, Ann Miller, Raymond Middleton, Harry Stockwell.
New York run: Alvin Theatre, Aug. 28, 1939; 120 performances.

1941. CRAZY WITH THE HEAT

Revue with sketches by Arthur Sheekman, Mack Davis, Max Liebman, etc.; music & lyrics mostly by Irvin Graham. Produced & directed by Kurt Kasznar; dances by Catherine Littlefield.

Cast included: Luella Gear, Gracie Barrie, Marie Nash, Betty Kean, Carl Randall, Richard Kollmar, Carlos Ramirez, Harold Gary.
New York run: 44th St. Theatre, Jan. 14, 1941; 99 performances.

1942. PRIORITIES OF 1942

Variety show. Produced by Clifford Fischer; dances by Marjery Fielding.

Cast included: Lou Holtz, Phil Baker, Paul Draper, Joan Merrill, Hazel Scott, Gene Sheldon, Al Kelly.

New York run: 46th St. Theatre, March 12, 1942; 354 performances.

1943. MY DEAR PUBLIC

Musical comedy with book & lyrics by Irving Caesar; music by Gerald Marks. Produced by Irving Caesar; directed by Edgar MacGregor; dances by Felice Sorel, Henry LeTang.

Cast included: David Burns, Ethel Shutta, Georgie Tapps, Nanette Fabray, Al Kelly.

New York run: 46th St. Theatre, Sept. 9, 1943; 45 performances.

1944. STAR AND GARTER

Revue with sketches by H. I. Phillips, etc.; music by Jerry Seelen, Irving Berlin, Harold Arlen, Jimmy McHugh, etc.; lyrics by Lester Lee, Irving Berlin, Johnny Mercer, Al Dubin, etc. Produced by Michael Todd; directed by Hassard Short; dances by Al White, Jr.

Cast included: Lois Andrews, Georgia Sothern, Chaz Chase, Al Kelly, Carrie Finnell.

NOTE: Mr. Howard and the above-mentioned cast members were in the Chicago company of the revue, which played at the Blackstone Theatre for seven weeks beginning Nov. 16, 1944. The original Broadway production opened at the Music Box, June 24, 1942, and ran 609 performances.

1945. THE PASSING SHOW

Revue with sketches uncredited; music & lyrics by Ross Thomas, Will Morrissey, Irving Actman, etc. Produced by the Messrs. Shubert; directed by Russell Mack; dances by Carl Randall, Joe Crosby, Mme. Kamarova.

Cast included: Sue Ryan, Bobby Morris, Richard Buckley, Al Kelly.

New York run: None.

NOTE: Revue opened at the Bushnell Memorial, Hartford, Nov. 9, 1945, and closed at the Erlanger Theatre, Chicago, Feb. 17, 1946.

1948. SALLY

Musical comedy with book by Guy Bolton & P. G. Wodehouse; music by Jerome Kern; lyrics by P. G. Wodehouse. Produced by Hunt Stromberg Jr. & William Berney; directed by Billy Gilbert; dances by Richard Barstow.

Cast included: Bambi Linn, Jack Goode, Bibi Osterwald, Robert Shackleton.

New York run: Martin Beck Theatre, May 6, 1948; 36 performances.

1948. ALONG FIFTH AVENUE

Revue with sketches by Charles Sherman, Hal Block, Mel Tolkin, Max Liebman, Nat Hiken; music by Gordon Jenkins; lyrics by Tom Adair. Produced by Arthur Lesser; directed by Charles Friedman; dances by Robert Sidney.

Cast included: Nancy Walker, Joyce Matthews, Carol Bruce, Hank Ladd, Viola Essen, Donald Richards, Virginia Gorski (Virginia Gibson), George S. Irving.

NOTE: Mr. Howard appeared in this revue only during its tryout at the Shubert Theatre, New Haven, beginning Nov. 24, 1948. Because of illness he was replaced by Jackie Gleason. The production then opened on Broadway at the Broadhurst Theatre, Jan. 13, 1949, and ran for 180 performances.

Bert Lahr

1927. HARRY DELMAR'S REVELS

Revue with sketches by William K. Wells; music by James Monaco, Jesse Greer, Lester Lee; lyrics by Billy Rose, Ballard MacDonald. Produced by Samuel Baerwitz & Harry Delmar; directed by Harry Delmar.

Cast included: Winnie Lightner, Frank Fay, Patsy Kelly, E. Mercedes.
New York run: Shubert Theatre, Nov. 28, 1927; 114 performances.

1928. HOLD EVERYTHING!

Musical comedy with book by B. G. DeSylva & John McGowan; music by Ray Henderson; lyrics by B. G. DeSylva & Lew Brown. Produced by Alex A. Aarons & Vinton Freedley; director uncredited; dances by Jack Haskell, Sam Rose.

Cast included: Victor Moore, Jack Whiting, Ona Munson, Nina Olivette, Betty Compton.
New York run: Broadhurst Theatre, Oct. 10, 1928; 413 performances.

1930. FLYING HIGH

Musical comedy with book by B. G. DeSylva & John McGowan; music by Ray Henderson; lyrics by B. G. DeSylva & Lew Brown. Produced by George White; directed by George White, Edward Clark Lilley; dances by George White, Bobby Connolly.

Cast included: Oscar Shaw, Grace Brinkley, Kate Smith, Russ Brown.
New York run: Apollo Theatre, March 3, 1930; 357 performances.

1932. HOT-CHA!

Musical comedy with book by Lew Brown, Ray Henderson, Mark Hellinger; music by Ray Henderson; lyrics by Lew Brown. Produced by Florenz Ziegfeld; directed by Edgar MacGregor, Edward Clarke Lilley; dances by Bobby Connolly.

Cast included: Lupe Velez, Buddy Rogers, Marjorie White, Lynne Overman, June Knight, June MacCloy, Velez & Yolanda, Antonio & Renée DeMarco, Eleanor Powell, Rose Louise (Gypse Rose Lee).
New York run: Ziegfeld Theatre, March 8, 1932; 119 performances.

1932. GEORGE WHITE'S MUSIC HALL VARIETIES

Revue with sketches by William K. Wells, George White; music by Harold Arlen, Sam H. Stept, etc.; lyrics by Irving Caesar, Herb Magidson, etc. Presented & directed by George White; dances by Russell Markert.

Cast included: Harry Richman, Lili Damita, Eleanor Powell, Loomis Sisters, Betty Kean, Willie & Eugene Howard (added Jan. 2, 1933).
New York run: Casino Theatre, Nov. 22, 1932; 72 performances.

1934. LIFE BEGINS AT 8:40

Revue with sketches by David Freedman, Allan Baxter, etc.; music by Harold Arlen; lyrics by Ira Gershwin & E. Y. Harburg. Produced by the Messrs. Shubert; directed by John Murray Anderson, Philip Loeb; dances by Robert Alton, Charles Weidman.

Cast included: Ray Bolger, Luella Gear, Frances Williams, Brian Donlevy, Dixie Dunbar.
New York run: Winter Garden, Aug. 27, 1934; 237 performances.

1935. GEORGE WHITE'S SCANDALS (12th Edition)

Revue with sketches by William K. Wells, George White, Howard Shiebler; music by Ray Henderson; lyrics by Jack Yellen. Produced & directed by George White; dances by Russell Markert.

Cast included: Willie & Eugene Howard, Rudy Vallee, Gracie Barrie, Cliff Edwards, Jane Cooper, Sam, Ted & Ray, Hal Forde.
New York run: New Amsterdam Theatre, Dec. 25, 1935; 110 performances.

1936. THE SHOW IS ON

Revue with sketches by David Freedman, Moss Hart; music mostly by Vernon Duke; lyrics mostly by Ted Fetter. Produced by the Messrs. Shubert; directed by Vincente Minnelli, Edward Clarke Lilley; dances by Robert Alton.

Cast included: Beatrice Lillie, Reginald Gardiner, Mitzi Mayfair, Paul Haakon, Gracie Barrie, Charles Walters, Vera Allen, Jack McCauley.
New York run: Winter Garden, Dec. 25, 1936; 237 performances.

1939. Du BARRY WAS A LADY

Musical comedy with book by Herbert Fields & B. G. DeSylva; music & lyrics by Cole Porter. Produced by B. G. DeSylva; directed by Edgar MacGregor; dances by Robert Alton.

Cast included: Ethel Merman, Betty Grable, Benny Baker, Ronald Graham, Charles Walters.
New York run: 46th St. Theatre, Dec. 6, 1939; 408 performances.

1944. SEVEN LIVELY ARTS

Revue with sketches by Moss Hart, Ben Hecht; music & lyrics by Cole Porter. Produced by Billy Rose; directed by Hassard Short, Philip Loeb; dances by Jack Donohue, Anton Dolin.

Cast included: Beatrice Lillie, Benny Goodman, Alicia Markova, Anton Dolin, Doc Rockwell, Nan Wynn, Jere McMahon, Paula Bane, Billie Worth, Bill Tabbert, Dolores Gray, Albert Carroll, Dennie Moore, Helen Gallagher, Red Norvo, Teddy Wilson.
New York run: Ziegfeld Theatre, Dec. 7, 1944; 183 performances.

1946. BURLESQUE

Play by George M. Watters & Arthur Hopkins. Produced by Jean Dalrymple; directed by Arthur Hopkins; burlesque sequence directed by Bert Lahr; dances by Billy Holbrook.

Cast included: Jean Parker, Ross Hertz, Joyce Matthews.
New York run: Belasco Theatre, Dec. 25, 1946; 439 performances.

1949. MAKE MINE MANHATTAN

Revue with sketches by Arnold B. Horwitt; music by Richard Lewine; lyrics by Arnold B. Horwitt. Produced by Joseph M. Hyman; directed by Hassard Short; dances by Lee Sherman.

Cast included: David Burns, Eric Brotherson, Lou Wills, Jr., Bob Fosse, Mary Ann Niles, Sheila Bond, Jack Albertson.
NOTE: Mr. Lahr and the above-mentioned cast members were in the touring company of the revue, which opened at the Shubert Theatre, Boston, Jan. 10, 1949, and closed at the Blackstone Theatre, Chicago, March 14, 1949. The original Broadway production opened at the Broadhurst Theatre, Jan. 15, 1948, and ran for 429 performances.

1951. TWO ON THE AISLE

Revue with sketches by Nat Hiken, William Friedberg, Betty Comden & Adolph Green; music by Jule Styne; lyrics by Betty Comden & Adolph Green. Produced by Arthur Lesser; directed by Abe Burrows; dances by Ted Cappy, Ruthanna Boris.

Cast included: Dolores Gray, Elliott Reid, Colette Marchand, Stanley Prager, J. C. McCord, Larry Laurence (Enzo Stuarti).
New York run: Mark Hellinger Theatre, July 19, 1951; 281 performances.

1956. WAITING FOR GODOT

Play by Samuel Beckett. Produced by Michael Myerberg; directed by Herbert Berghof.

Cast included: E. G. Marshall, Kurt Kasznar, Alvin Epstein.
New York run: John Golden Theatre, April 19, 1956; 59 performances.

1957. HOTEL PARADISO

Play by Georges Feydeau & Maurice Desvallières. Produced by Richard Myers, Julius Fleischmann & Bowden, Barr & Bullock; directed by Peter Glenville.

Cast included: Angela Lansbury, John Emery, Carleton Carpenter, Douglas Byng, Vera Pearce, James Coco.
New York run: Henry Miller's Theatre, April 11, 1957; 108 performances.

1959. THE GIRLS AGAINST THE BOYS

Revue with sketches & lyrics by Arnold B. Horwitt; music by Richard Lewine. Produced by Albert Selden; directed by Aaron Ruben; dances by Boris Runanin.

Cast included: Nancy Walker, Shelley Berman, Dick Van Dyke, Joy Nichols, Imelda De Martin, Martin Charnin, Cy Young.
New York run: Alvin Theatre, Nov. 2, 1959; 16 performances.

1960. A MIDSUMMER NIGHT'S DREAM

Play by William Shakespeare. Produced by the American Shakespeare Festival Co.; directed by Jack Landau; dances by George Balanchine.

Cast included: Douglas Watson, William Hickey, Will Geer, Clayton Corzatte, Richard Waring, Margaret Phillips, Alexandra Berlin, Patrick Hines.
New York run: None.
NOTE: Production opened at the Colonial Theatre, Boston, Sept. 26, 1960, and closed at the National Theatre, Washington, D.C., Feb. 25, 1961.

1962. THE BEAUTY PART

Play by S. J. Perelman. Produced by Michael Ellis; directed by Noel Willman.

Cast included: Alice Ghostley, Larry Hagman, David Doyle, Bernie West, William LeMassena, Charlotte Rae, Arnold Soboloff.
New York run: Music Box, Dec. 26, 1962; 85 performances.

1964. FOXY

Musical comedy with book by Ian McLellan Hunter & Ring Lardner, Jr.; music by Robert Emmett Dolan; lyrics by Johnny Mercer. Produced by David Merrick; directed by Robert Lewis; dances by Jack Cole.

Cast included: Larry Blyden, Julienne Marie, Cathryn Damon, Gerald Hiken, John Davidson, Robert H. Harris.
New York run: Ziegfeld Theatre, Feb. 16, 1964; 72 performances.

Beatrice Lillie

1914. NOT LIKELY!

Revue with sketches by George Grossmith, Jr., Cosmo Gordon-Lennox; music & lyrics by miscellaneous writers. Produced by André Charlot; directed by George Grossmith, Jr., Cosmo Gordon-Lennox; dances by J. W. Jackson.

Cast included: Eddie Cantor, Teddie Gerard, George Grossmith Jr., Lee White, Clay Smith, Phyllis Monkman.
London run: Alhambra Theatre, May 4, 1914; 305 performances.
NOTE: Miss Lillie joined the cast in October 1914.

1915. 5064 GERRARD

Revue with sketches by Cosmo Gordon-Lennox, Robert Hale, C. H. Bovill; music & lyrics by miscellaneous writers. Produced & directed by André Charlot; dances by J. W. Jackson.

Cast included: Robert Hale, Lee White, Clay Smith, Phyllis Monkman, Oscar Shaw, Lilian Davies.
London run: Alhambra Theatre, March 19, 1915; 194 performances.

1915. NOW'S THE TIME

Revue with sketches by Cosmo Gordon-Lennox, C. H. Bovill; music & lyrics by miscellaneous writers. Produced by André Charlot; directed by Herbert Bryan; dances by J. W. Jackson.

Cast included: Lee White, Clay Smith, Phyllis Monkman, Gene Gerrard.
London run: Alhambra Theatre, Oct. 13, 1915; 147 performances.

1916. SAMPLES

Revue with sketches by Harry Grattan; music & lyrics by miscellaneous writers. Produced & directed by André Charlot; dances by George Shurley.

Cast included: Melville Gideon, Bert Coote, Terry Twins, Davy Burnaby (added March 1916).
London run: Playhouse, Nov. 30, 1915; 242 performances.
NOTE: Miss Lillie replaced Mabel Russell in March 1916 when the revue moved to the Vaudeville Theatre.

1916. SOME

Revue with sketches by Harry Grattan; music mostly by James W. Tate; lyrics mostly by Clifford Harris, Arthur Valentine. Produced by André Charlot; directed by Harry Grattan; dances by Frank Gordon.

Cast included: Lee White, Clay Smith, Rebla, Gene Gerrard, Gertrude Lawrence, Guy LeFeuvre.
London run: Vaudeville Theatre, June 29, 1916; 273 performances.

1917. CHEEP

Revue with sketches by Harry Grattan; music & lyrics by miscellaneous writers. Produced by André Charlot; directed by Harry Grattan; dances by Frank Gordon.

Cast included: Lee White, Clay Smith, Teddie Gerard, Walter Williams, Guy LeFeuvre.
London run: Vaudeville Theatre, April 26, 1917; 483 performances.

1918. TABS

Revue with sketches by Ronald Jeans; music by Ivor Novello; lyrics by Ronald Jeans. Produced by André Charlot; directed by Ronald Jeans; dances by Gwladys Dillon.

Cast included: Gertrude Lawrence, Alfred Austin, Guy LeFeuvre, Walter Williams, Odette Myrtil.
London run: Vaudeville Theatre, May 15, 1918; 268 performances.
NOTE: Miss Lillie was replaced by Gertrude Lawrence for two months during run.

1919. OH, JOY!

Musical comedy with book by P. G. Wodehouse & Guy Bolton; music by Jerome Kern; lyrics by P. G. Wodehouse. Produced by George Grosssmith & Edward Laurillard; directed by Austen Hurgon; dances by Harry French, Hylda Lewis.

Cast included: Tom Powers, Isabel Jeans, Billy Leonard, Hal Gordon.
London run: Kingsway Theatre, Jan. 27, 1919; 167 performances.

1919. BRAN PIE

Revue with sketches by Ronald Jeans, John Hastings Turner, William Anthony McGuire, Jack Hulbert; music by Philip Braham, J. Russel Robinson, Pete Wendling; lyrics by Jack Hulbert, Al Kendall, Edgar Leslie, Bert Kalmar. Produced by André Charlot; directed by Herbert Bryan; dances by Larry Ceballos.

Cast included: Jack Hulbert, Jack Barker, Odette Myrtil.
London run: Prince of Wales Theatre, Aug. 28, 1919; 414 performances.
NOTE: Miss Lillie left the cast before June 1920.

1921. UP IN MABEL'S ROOM

Play by Wilson Collison & Otto Harbach.

Cast included: Charles Hawtry, Isobel Elsom, Doris Kendall, Stanely Brett.
London run: Playhouse; April 6, 1921; 37 performances.

1921. NOW AND THEN

Revue with sketches by John Hastings Turner; music by Philip Braham; lyrics by Reginald Arkell. Produced by André Charlot; directed by Dion Titheradge.

Cast included: Joyce Barbour, Roy Royston, George Graves, Miles Malleson, Fred Hearne.
London run: Vaudeville Theatre, Sept. 17, 1921; 76 performances.
NOTE: Miss Lillie joined the cast Oct. 1921.

1921. POT LUCK

Revue with sketches by Ronald Jeans, Dion Titheradge; music by Philip Braham; lyrics by R.P. Weston & Bert Lee. Produced by André Charlot; directed by Dion Titheradge.

Cast included: Jack Hulbert, Herbert Mundin, Mary Leigh, Norah Blaney, Gus Sharland.
London run: Vaudeville Theatre, Dec. 24, 1921; 284 performances.

1922. A TO Z

Revue with sketches by Dion Titheradge, Ronald Jeans; music by Ivor Novello; lyrics by Dion Titheradge, Ronald Jeans. Produced & directed by André Charlot.

Cast included: Jack Buchanan, Helen & Josephine Trix, Herbert Mundin, Teddie Gerard, Douglas Furber, Maisie Gay.
London run: Prince of Wales Theatre, Oct. 12, 1921; 428 performances.
NOTE: Miss Lillie replaced Gertrude Lawrence Sept. 1922.

1922. THE NINE O'CLOCK REVUE

Revue with sketches by Harold Simpson, Morris Harvey; music mostly by Muriel Lillie; lyrics mostly by Arthur Weigall, Harold Simpson. Produced by J. L. Davies; directed by Dion Titheradge; dances by George Shurley.

Cast included: Morris Harvey, Irene Browne, Mimi Crawford.
London run: Little Theatre, Oct. 25, 1922; 385 performances.

1924. ANDRÉ CHARLOT'S REVUE OF 1924

Revue with sketches by Ronald Jeans, Dion Titheradge; music by Philip Braham, Noël Coward, Ivor Novello; lyrics by Ronald Jeans, Douglas Furber. Produced by Arch Selwyn; directed by André Charlot; dances by David Bennett.

Cast included: Jack Buchanan, Gertrude Lawrence, Douglas Furber, Herbert Mundin, Fred Leslie, Jessie Matthews, Constance Carpenter.
New York run: Times Square Theatre, Jan. 9, 1924; 298 performances.

1925. CHARLOT'S REVUE (2nd Edition)

Revue with sketches by Dion Titheradge, Ronald Jeans; music by Philip Braham, Ivor Novello, Noël Coward; lyrics by Douglas Furber, Noël Coward, Ronald Jeans, Dion Titheradge. Produced by André Charlot; directed by Dion Titheradge.

Cast included: Gertrude Lawrence, Herbert Mundin, Jessie Matthews, Robert Hobbs, Effie Atherton.
London run: Prince of Wales Theatre, March 3, 1925.
NOTE: This was the 2nd Edition of a revue that had opened Sept. 23, 1924, and which had a total run of 513 performances. Miss Lillie left the cast in July 1925.

1925. THE CHARLOT REVUE OF 1926

Revue with sketches by Douglas Furber, Ronald Jeans; music by Philip Braham, Noël Coward, Ivor Novello; lyrics by Douglas Furber, Noël Coward, Dion Titherage. Produced by Arch Selwyn; directed by André Charlot; dances by Jack Buchanan.

Cast included: Jack Buchanan, Gertrude Lawrence, Betty Stockfield, Herbert Mundin, Constance Carpenter.
New York run: Selwyn Theatre, Nov. 10, 1925; 138 performances.

1926. OH, PLEASE!

Musical comedy with book by Otto Harbach & Anne Caldwell; music by Vincent Youmans; lyrics by Anne Caldwell. Produced by Charles B. Dillingham; directed by Hassard Short; dances by David Bennett.

Cast included: Charles Winninger, Charles Purcell, Kitty Kelly, Helen Broderick, Charles Columbus, Nick Long, Jr.
New York run: Fulton Theatre, Dec. 17, 1926; 75 performances.

1928. SHE'S MY BABY

Musical comedy with book by Guy Bolton, Bert Kalmar, & Harry Ruby; music by Richard Rodgers; lyrics by Lorenz Hart. Produced by Charles B. Dillingham; directed by Edward Royce; dances by Mary Read.

Cast included: Clifton Webb, Irene Dunne, Jack Whiting, Nick Long, Jr., William Frawley.
New York run: Globe Theatre, Jan. 3, 1928; 71 performances.

1928. THIS YEAR OF GRACE

Revue with sketches, music & lyrics by Noël Coward. Produced by Arch Selwyn; directed by Frank Collins; dances by Max Rivers.

Cast included: Noël Coward, Florence Desmond, Queenie Leonard, Madeline Gibson, Billy Milton.
New York run: Selwyn Theatre, Nov. 7, 1928; 158 performances.

1930. CHARLOT'S MASQUERADE

Revue with sketches by Ronald Jeans; music by Reg Casson, etc.; lyrics by Rowland Leigh. Produced by B. A. Meyer; directed by André Charlot, Auriol Lee, Ronald Jeans; dances by Harland Dixon, Quentin Tod.

Cast included: Henry Kendall, J. H. Roberts, Constance Carpenter, Florence Desmond, Quentin Tod, Anton Dolin, Dora Vadimova.
London run: Cambridge Theatre, Sept. 4, 1930.

1931. THE THIRD LITTLE SHOW

Revue with sketches by Noël Coward, S. J. Perelman, Peter Spencer, Marc Connelly; music mostly by Ned Lehak; lyrics mostly by Edward Eliscu. Produced by Dwight Deere Wiman with Tom Weatherly; directed by Alexander Leftwich; dances by Dave Gould.

Cast included: Ernest Truex, Walter O'Keefe, Jerry Norris, Carl Randall, Constance Carpenter, Edward Arnold, Gertrude McDonald.
New York run: Music Box, June 1, 1931; 136 performances.

1932. TOO TRUE TO BE GOOD

Play by George Bernard Shaw. Produced by the Theatre Guild; directed by Leslie Banks.

Cast included: Hope Williams, Hugh Sinclair, Ernest Cossart, Leo G. Carroll, Claude Rains.
New York run: Guild Theatre, April 4, 1932; 57 performances.

1932. WALK A LITTLE FASTER

Revue with sketches by S. J. Perelman, Robert MacGunigle; music by Vernon Duke; lyrics by E. Y. Harburg. Produced by Courtney Burr; directed by Monty Woolley; dances by Albertina Rasch.

Cast included: Bobby Clark & Paul McCullough, John Hundley, Evelyn Hoey, Donald Burr, Dorothy McNulty (Penny Singleton), Dave & Dorothy Fitzgibbon.
New York run: St. James Theatre, Dec. 7, 1932; 119 performances.

1933. PLEASE

Revue with sketches by Dion Titheradge, Robert MacGunigle; music mostly by Vivian Ellis, Austin Croom-Johnston; lyrics mostly by Dion Titheradge. Produced by André Charlot; directed by Dion Titheradge; dances by Jack Donohue, Alanova.

Cast included: Lupino Lane, Kenneth Carten, George Benson.
London run: Savoy Theatre, Nov. 16, 1933; 108 performances.

1935. AT HOME ABROAD

Revue with sketches by Howard Dietz, Dion Titheradge, Raymond Knight, Marc Connelly; music by Arthur Schwartz; lyrics by Howard Dietz. Produced by the Messrs. Shubert; directed by Vincente Minnelli, Thomas Mitchell; dances by Gene Snyder, Harry Losee.

Cast included: Ethel Waters, Eleanor Powell, Reginald Gardiner, Herb Williams, Paul Haakon, Eddie Foy, Jr., Vera Allen, Nina Whitney, John Payne.
New York run: Winter Garden, Sept. 19, 1935; 198 performances.

1936. THE SHOW IS ON

Revue with sketches by David Freedman, Moss Hart; music mostly by Vernon Duke; lyrics mostly by Ted Fetter. Produced by the Messrs. Shubert; directed by Vincente Minnelli, Edward Clarke Lilley; dances by Robert Alton.

Cast included: Bert Lahr, Reginald Gardiner, Mitzi Mayfair, Paul Haakon, Gracie Barrie, Charles Walters, Vera Allen, Jack McCauley.
New York run: Winter Garden, Dec. 25, 1936; 237 performances.

1938. HAPPY RETURNS

Revue with sketches by Moss Hart, David Freedman; music mostly by Arthur Johnston; lyrics mostly by Ian Grant. Produced by Charles B. Cochran; directed by Edward Duryea Dowling; dances by Buddy Bradley.

Cast included: Bud Flanagan & Chesney Allen, Patricia Burke, Constance Carpenter, Phyllis Stanley, Gerald Nodin.
London run: Adelphi Theatre, May 19, 1938.

1939. SET TO MUSIC

Revue with sketches, music & lyrics by Noël Coward. Produced by John C. Wilson; directed by Noël Coward.

Cast included: Richard Haydn, Eva Ortega, Hugh French, Penelope Dudley Ward.
New York run: Music Box, Jan. 18, 1939; 129 performances.

1939. ALL CLEAR

Revue with sketches by Harold French, Arthur Macrae; music mostly by Richard Addinsell; lyrics mostly by Arthur Macrae. Produced by H. M. Tennent; directed by Harold French; dances by Buddy Bradley.

Cast included: Bobby Howes, Fred Emney, Adele Dixon, Hugh French.
London run: Queen's Theatre, Dec. 20, 1939; 162 performances.

1942. BIG TOP

Revue with sketches by Herbert Farjeon; music mostly by Geoffrey Wright; lyrics mostly by Herbert Farjeon. Produced by Charles B. Cochran; directed by Frank Collins; dances by Buddy Bradley.

Cast included: Cyril Ritchard, Fred Emney, Madge Elliott, Patricia Burke, Charles Hickman.
London run: His Majesty's Theatre, May 8, 1942; 139 performances.

1944. SEVEN LIVELY ARTS

Revue with sketches mostly by Moss Hart, Ben Hecht; music & lyrics by Cole Porter. Produced by Billy Rose; directed by Hassard Short, Philip Loeb; dances by Jack Donohue, Anton Dolin.

Cast included: Bert Lahr, Benny Goodman, Alicia Markova, Anton Dolin, Doc Rockwell, Nan Wynn, Jere McMahon, Paula Bane, Billie Worth, Bill Tabbert, Dolores Gray, Albert Carroll, Dennie Moore, Helen Gallagher, Red Norvo, Teddy Wilson.
New York run: Ziegfeld Theatre, Dec. 7, 1944; 183 performances.

1946. BETTER LATE

Revue with sketches by Leslie Julian Jones; music mostly by Geoffrey Wright; lyrics mostly by Nicholas Phipps. Produced by James Lavall, Leslie Julian Jones, Anthony Wickham; directed by Norman Marshall; dances by William Chappell.

Cast included: Walter Crisham, Murray Matheson, Joan Swinstead.
London run: Garrick Theatre, April 24, 1946; 211 performances.

1948. INSIDE U.S.A.

Revue with sketches by Arnold Auerbach, Moss Hart, Arnold Horwitt; music by Arthur Schwartz; lyrics by Howard Dietz. Produced by Arthur Schwartz; directed by Robert H. Gordon; dances by Helen Tamiris.

Cast included: Jack Haley, John Tyers, Valerie Bettis, Herb Shriner, Eric Victor, Louis Nye, Carl Reiner, Thelma Carpenter, Tally Beatty, Estelle Loring, Rod Alexander, Jack Cassidy.
New York run: New Century Theatre, April 30, 1948; 399 performances.

1952. AN EVENING WITH BEATRICE LILLIE

Revue with sketches by Robert MacGunigle, Howard Simpson; music & lyrics by miscellaneous writers. Produced & directed by Edward Duryea Dowling.

Cast included: Reginald Gardiner, Xenia Banks, Florence Bray, John Philip.
New York run: Booth Theatre, Oct. 2, 1952; 276 performances.
NOTE: Same basic revue opened in London at the Globe Theatre, Nov. 24, 1954, and ran for 195 performances. The cast included Leslie Bricusse, Constance Carpenter, Frances Clare, and John Philip.

1957. ZIEGFELD FOLLIES

Revue with sketches, music & lyrics mostly by Richard Myers & Jack Lawrence, Dean Fuller & Marshall Barer. Produced by Mark Kroll & Charles Conaway; directed by John Kennedy; dances by Frank Wagner.

Cast included: Billy DeWolfe, Harold Lang, Jane Morgan, Helen Wood, Carol Lawrence, Paula Wayne.
New York run: Winter Garden, March 1, 1957; 123 performances.

1958. AUNTIE MAME

Play by Jerome Lawrence & Robert E. Lee. Produced by Robert Fryer & Lawrence Carr; directed by Morton DaCosta.

Cast included: Polly Rowles, Florence MacMichael, Walter Greaza, Robert Smith, James Monk, Dennis Joel, Ray Fulmer, Spivy.
New York run: Broadhurst Theatre, Oct. 31, 1956; 637 performances.
NOTE: Miss Lillie appeared with the above-mentioned cast members for the play's final 32 performances on Broadway. She succeeded Greer Garson who had succeeded Rosalind Russell. With Miss Lillie starred, the play opened in London at the Adelphi Theatre, Sept. 10, 1958, and ran 301 performances. The cast included Florence Desmond, Rosamund Greenwood, Donald Stewart, Geoffrey Toone, John Hall, Dinsdale Landen, and Natalie Lynn.

1964. HIGH SPIRITS

Musical comedy with book, music & lyrics by Hugh Martin & Timothy Gray. Produced by Lester Osterman, Robert Fletcher, Richard Horner; directed by Noël Coward, Gower Champion (uncredited); dances by Danny Daniels.

Cast included: Tammy Grimes, Edward Woodward, Louise Troy, Carol Arthur, Margaret Hall.
New York run: Alvin Theatre, April 7, 1964; 375 performances.

Victor Moore

1896. ROSEMARY

Play by Lewis N. Parker & Murray Carson. Produced by Charles Frohman; directed by Joseph Humphreys.

Cast included: John Drew, Maude Adams, Ethel Barrymore, Arthur Byron.
New York run: Empire Theatre, Aug. 31, 1896; 136 performances.

1897. SPIRITISME

Play by Victorien Sardou. Produced by Charles Frohman & Al Hayman; directed by Joseph Humphreys.

Cast included: Maurice Barrymore, Virginia Harned, Nelson Wheatcroft.
New York run: Knickerbocker Theatre, Feb. 22, 1897; 24 performances.

1906. FORTY-FIVE MINUTES FROM BROADWAY

Musical comedy with book, music & lyrics by George M. Cohan. Produced by Klaw & Erlanger; directed by George M. Cohan.

Cast included: Fay Templeton, Donald Brian, Lois Ewell, Julia Ralph.
New York run: New Amsterdam Theatre, Jan. 1, 1906; 90 performances.

1907. THE TALK OF NEW YORK

Musical comedy with book, music & lyrics by George M. Cohan. Produced by Cohan & Harris; directed by George M. Cohan.

Cast included: Stanley H. Forde, Saidie Harris, Jack Gardner, Gertrude Vanderbilt, Rosie Green, Emma Littlefield.
New York run: Knickerbocker Theatre, Dec. 3, 1907; 157 performances.

1911. THE HAPPIEST NIGHT OF HIS LIFE

Musical comedy with book & lyrics by Junie McCree & Sydney Rosenfeld; music by Albert Von Tilzer. Produced by H. H. Frazee & George W. Lederer; directed by George W. Lederer.

Cast included: Annabelle Whitford, Phil Ryley, Jack Henderson, Junie McCree, Gertrude Vanderbilt, Emma Littlefield.
New York run: Criterion Theatre, Feb. 29, 1911; 24 performances.

1911. SHORTY McCABE

Play by Owen Davis. Produced by H. H. Frazee & George W. Lederer; directed by Charles W. Lederer.

Cast included: Jean Galbraith, Maidel Turner, Emma Littlefield.
New York run: none.
NOTE: Play opened at the Cort Theatre, Chicago, where it ran a week, then toured.

1918. PATSY ON THE WING

Play by Edward Peple. Produced by Harrison Grey Fiske.

Cast included: Peggy O'Neil, Edgar Norton, J. H. Gilmour.
New York run: none.
NOTE: Play opened at the Broadway Theatre, Long Branch, N.J., June 24, 1918, then played Chicago and toured.

1919. SEE YOU LATER

Musical comedy with book by P. G. Wodehouse & Guy Bolton; lyrics by P. G. Wodehouse; music by Jean Schwartz. Produced by William Elliott, F. Ray Comstock, & Morris Gest; directed by Robert Milton; dances by Julian Alfred.

Cast included: T. Roy Barnes, Winona Winter, Frances Cameron.
New York run: none.
NOTE: Musical opened at the LaSalle Theatre, Chicago, on Jan. 16, 1919, then toured.

1925. EASY COME, EASY GO

Play by Owen Davis. Produced by Lewis & Gordon; directed by Priestley Morrison.

Cast included: Otto Kruger, Edward Arnold, Betty Garde, Nan Sunderland, Vaughn DeLeath.
New York run: George M. Cohan Theatre, Oct. 26, 1925; 180 performances.

1926. OH, KAY!

Musical comedy with book by Guy Bolton & P. G. Wodehouse; music by George Gershwin; lyrics by Ira Gershwin. Produced by Alex A. Aarons & Vinton Freedley; directed by John Harwood; dances by Sammy Lee.

Cast included: Gertrude Lawrence, Oscar Shaw, Harland Dixon, Fairbanks Twins, Betty Compton, Constance Carpenter, Gerald Oliver Smith, Frank Gardiner, Sascha Beaumont.

New York run: Imperial Theatre, Nov. 8, 1926; 257 performances.

1927. ALLEZ-OOP!

Revue with sketches by J. P. McEvoy; music by Phil Charig, Richard Myers; lyrics by Leo Robin. Produced & directed by Carl Hemmer.

Cast included: Charles Butterworth, Bobby Watson, Evelyn Bennett, Esther Howard, Madeleine Fairbanks.

New York run: Earl Carroll Theatre, Aug. 2, 1927; 120 performances.

NOTE: Mr. Moore left the cast in Oct. 1927.

1927. FUNNY FACE

Musical comedy with book by Fred Thompson & Paul Gerard Smith; music by George Gershwin; lyrics by Ira Gershwin. Produced by Alex A. Aarons & Vinton Freedley; directed by Edgar MacGregor; dances by Bobby Connolly.

Cast included: Fred & Adele Astaire, William Kent, Allen Kearns, Betty Compton, Dorothy Jordan, Gertrude McDonald, Earl Hampton.

New York run: Alvin Theatre, Nov. 22, 1927; 250 performances.

1928. HOLD EVERYTHING!

Musical comedy with book by B. G. DeSylva & John McGowan; music by Ray Henderson; lyrics by B. G. DeSylva & Lew Brown. Produced by Alex A. Aarons & Vinton Freedley; director uncredited; dances by Jack Haskell, Sam Rose.

Cast included: Bert Lahr, Jack Whiting, Ona Munson, Nina Olivette, Betty Compton.

New York run: Broadhurst Theatre, Oct. 10, 1928; 413 performances.

1929. HEADS UP!

Musical comedy with book by John McGowan & Paul Gerard Smith; music by Richard Rodgers; lyrics by Lorenz Hart. Produced by Alex A. Aarons & Vinton Freedley; directed by George Hale; dances by George Hale.

Cast included: Betty Starbuck, Barbara Newberry, Jack Whiting, Ray Bolger, Robert Gleckler.

New York run: Alvin Theatre, Nov. 11, 1929; 144 performances.

1930. PRINCESS CHARMING

Musical comedy with book by Jack Donahue; music by Albert Sirmay & Arthur Schwartz; lyrics by Arthur Swanstrom. Produced by Bobby Connolly & Arthur Swanstrom; directed by Bobby Connolly, Edward Clarke Lilley; dances by Albertina Rasch.

Cast included: Evelyn Herbert, Robert Halliday, George Grossmith, Jeanne Aubert, Douglas Dumbrille, Howard St. John.
New York run: Imperial Theatre, Oct. 13, 1930; 56 performances.

1931. SHE LIVED NEXT TO THE FIREHOUSE

Play by William Grew & Harry Delf. Produced by L. Lawrence Weber; directed by William B. Friedlander.

Cast included: Ara Gerald, Ralph Hertz, William Frawley.
New York run: Longacre Theatre, Feb. 10, 1931; 24 performances.

1931. OF THEE I SING

Musical comedy with book by George S. Kaufman & Morrie Ryskind; music by George Gershwin; lyrics by Ira Gershwin. Produced by Sam H. Harris; directed by George S. Kaufman; dances by George Hale.

Cast included: William Gaxton, Lois Moran, Grace Brinkley, Dudley Clements, Florenz Ames, George Murphy, Ralph Riggs.
New York run: Music Box, Dec. 26, 1931; 441 performances.

1933. LET 'EM EAT CAKE

Musical comedy with book by George S. Kaufman & Morrie Ryskind; music by George Gershwin; lyrics by Ira Gershwin. Produced by Sam H. Harris; directed by George S. Kaufman; dances by Eugene Van Grona, Ned McGurn.

Cast included: William Gaxton, Lois Moran, Dudley Clements, Philip Loeb, Florenz Ames.
New York run: Imperial Theatre, Oct. 21, 1933; 90 performances.

1934. ANYTHING GOES

Musical comedy with book by Guy Bolton & P. G. Wodehouse revised by Howard Lindsay & Russel Crouse; music & lyrics by Cole Porter. Produced by Vinton Freedley; directed by Howard Lindsay; dances by Robert Alton.

Cast included: William Gaxton, Ethel Merman, Bettina Hall, Vivian Vance.
New York run: Alvin Theatre, Nov. 21, 1934; 420 performances.

1938. LEAVE IT TO ME!

Musical comedy with book by Bella & Sam Spewack; music & lyrics by Cole Porter. Produced by Vinton Freedley; directed by Sam Spewack; dances by Robert Alton.

Cast included: William Gaxton, Tamara, Sophie Tucker, Mary Martin, Gene Kelly, George Tobias.
New York run: Imperial Theatre, Nov. 9, 1938; 291 performances.

1940. LOUISIANA PURCHASE

Musical comedy with book by Morrie Ryskind; music & lyrics by Irving Berlin. Produced by B. G. DeSylva; directed by Edgar MacGregor; dances by Carl Randall, George Balanchine.

Cast included: William Gaxton, Vera Zorina, Irene Bordoni, Carol Bruce, Hugh Martin, Ralph Blane.
New York run: Imperial Theatre, May 28, 1940; 444 performances.

1942. KEEP 'EM LAUGHING

Variety show. Produced by Clifford Fischer & the Messrs. Shubert; directed by Clifford Fischer.

Cast included: William Gaxton, Paul & Grace Hartman, Hildegarde, Zero Mostel, Jack Cole.
New York run: 44th St. Theatre, April 24, 1942; 77 performances.

1945. HOLLYWOOD PINAFORE

Musical comedy with book & lyrics by George S. Kaufman; music by Arthur Sullivan. Produced by Max Gordon; directed by George S. Kaufman; dances by Douglas Coudy, Antony Tudor.

Cast included: William Gaxton, Shirley Booth, Annamary Dickey, Mary Wickes, Russ Brown, Viola Essen.
New York run: Alvin Theatre, May 31, 1945; 53 performances.

1946. NELLIE BLY

Musical comedy with book by Joseph Quinlan; music by James Van Heusen; lyrics by Johnny Burke. Produced by Nat Karson & Eddie Cantor; directed by Edgar MacGregor; dances by Lee Sherman.

Cast included: William Gaxton, Joy Hodges, Robert Strauss, Benay Venuta.
New York run: Adelphi Theatre, Jan. 21, 1946; 16 performances.

1953. ON BORROWED TIME

Play by Paul Osborn. Produced by Richard W. Krakeur & Randolph Hale; directed by Marshall Jamison.

Cast included: Beulah Bondi, Leo G. Carroll, David John Stollery, Kay Hammond.
New York run: 48th St. Theatre, Feb. 10, 1953; 78 performances.
NOTE: Mr. Moore previously appeared in the play at the Geary Theatre, San Francisco, for three weeks beginning July 11, 1938. That cast included Leona Roberts, Guy Bates Post, James West, and Myra Marsh.

1957. CAROUSEL

Musical play with book & lyrics by Oscar Hammerstein II; music by Richard Rodgers. Produced by the New York City Center Light Opera Co., Jean Dalrymple, director; directed by John Fearnley; dances by Agnes de Mille.

Cast included: Howard Keel, Barbara Cook, Pat Stanley, Kay Medford, Russell Nype, James Mitchell, Marie Powers, Bambi Linn, Patricia Birsh.
New York run: N.Y. City Center, Sept. 11, 1957; 24 performances.

Ed Wynn

1910. THE DEACON AND THE LADY

Musical comedy with book & lyrics by George Totten Smith; music by Alfred E. Aarons. Produced by Alfred E. Aarons & Louis Werba; directed by Alfred E. Aarons.

Cast included: Harry Kelly, Clara Palmer, Fletcher Norton.
New York run: New York Theatre, Oct. 4, 1910; 16 performances.

1914. ZIEGFELD FOLLIES (8th Edition)

Revue with sketches by George V. Hobart; music by Raymond Hubbell, David Stamper; lyrics by George V. Hobart, Gene Buck. Produced by Florenz Ziegfeld; directed by Leon Errol.

Cast included: Leon Errol, Ann Pennington, Bert Williams, Gloria Vanderbilt, Annette Kellerman, Vera Michelena.
New York run: New Amsterdam Theatre, June 1, 1914; 112 performances.

1915. ZIEGFELD FOLLIES (9th Edition)

Revue with sketches & lyrics by Channing Pollock, Rennold Wolf, Gene Buck; music by Louis A. Hirsch, David Stamper. Produced by Florenz Ziegfeld; directed by Julian Mitchell, Leon Errol.

Cast included: W.C. Fields, Ina Claire, Ann Pennington, Justine Johnstone, Mae Murray, Bernard Granville, Bert Williams, George White, Leon Errol, Carl Randall.
New York run: New Amsterdam Theatre, June 21, 1915; 104 performances.

1916. THE PASSING SHOW OF 1916

Revue with sketches & lyrics by Harold Atteridge; music by Sigmund Romberg, Otto Motzan. Produced by the Messrs. Shubert; directed by J.C. Huffman; dances by Allan K. Foster.

Cast included: James Hussey, Herman Timberg, Stella Hoban, Fred Walton.
New York run: Winter Garden, June 22, 1916; 140 performances.

1917. DOING OUR BIT

Revue with sketches & lyrics by Harold Atteridge; music by Sigmund Romberg, Herman Timberg. produced by the Messrs. Shubert; directed by J. C. Huffman; dances by Allan K. Foster.

Cast included: Ada Lewis, Herman Timberg, Duncan Sisters, Charles Judels, James J. Corbett, Frank Tinney.
New York run: Winter Garden, Oct. 18, 1917; 130 performances.
NOTE: Mr. Wynn left the cast after two months.

1917. OVER THE TOP

Revue with sketches by Philip Bartholomae, Harold Atteridge; music by Sigmund Romberg, Herman Timberg; lyrics by Charles J. Manning & Matthew Woodward. Produced by the Messrs. Shubert; directed by J. C. Huffman, Joseph Herbert; dances by Allan K. Foster.

Cast included: Fred & Adele Astaire, Justine Johnstone, Ted Lorraine, Mary Eaton, Craig Campbell, Joe Laurie.
New York run: 44th St. Roof Theatre, Nov. 28, 1917; 78 performances.
NOTE: Mr. Wynn replaced T. Roy Barnes in Dec. 1917.

1918. SOMETIME

Musical comedy with book & lyrics by Rida Johnson Young, Ed Wynn (uncredited); music by Rudolf Friml. Produced by Arthur Hammerstein; directed by Oscar Eagle; dances by Allan K. Foster.

Cast included: Francine Larrimore, Mae West, Frances Cameron, John Merkyl.
New York run: Shubert Theatre, Oct. 4, 1918; 283 performances.

1919. THE SHUBERT GAIETIES OF 1919

Revue with sketches by Edgar Smith, Harold Atteridge, Ed Wynn; music by Jean Schwartz; lyrics by Alfred Bryan. Produced by the Messrs. Shubert; directed by J. C. Huffman.

Cast included: Gilda Gray, William Kent, Ted Lorraine & Gladys Walton, Marjorie Gateson, Harry Fender, George Hassell, Marguerite Farrell.
New York run: 44th St. Theatre, July 17, 1919; 87 performances.
NOTE: The run was interrupted by the actors' strike early in August. Mr. Wynn was not in the cast when the show resumed its run in September.

1920. ED WYNN CARNIVAL

Variety show with dialogue, music & lyrics mostly by Ed Wynn. Produced by Ed Wynn & B. C. Whitney; directed by Ned Wayburn.

Cast included: Marion Davies, Earl Benham, Regal & Moore, Richie Ling.
New York run: New Amsterdam Theatre, April 5, 1920; 150 performances.

1921. THE PERFECT FOOL

Variety show with dialogue, music & lyrics by Ed Wynn. Produced by A. L. Erlanger & B. C. Whitney; directed by Julian Mitchell.

Cast included: Guy Robertson, Janet Velie, True Rice, Flo Newton.
New York run: George M. Cohan Theatre, Nov. 7, 1921; 275 performances.

1924. THE GRAB BAG

Variety show with dialogue, music & lyrics by Ed Wynn. Produced by Ed Wynn & A. L. Erlanger; directed by Julian Mitchell.

Cast included: Marion Fairbanks, Shaw & Lee, Janet Velie, Ralph Riggs.
New York run: Globe Theatre, Oct. 6, 1924; 184 performances.

1927. MANHATTAN MARY

Musical comedy with book by William K. Wells & George White; music by Ray Henderson; lyrics by B. G. DeSylva & Lew Brown. Produced & directed by George White.

Cast included: George White, Lou Holtz, Ona Munson, Paul Frawley, Harland Dixon, Doree Leslie, Mae Clarke, McCarthy Sisters.
New York run: Apollo Theatre, Sept. 26, 1927; 264 performances.

1930. SIMPLE SIMON

Musical comedy with book by Ed Wynn & Guy Bolton; music by Richard Rodgers; lyrics by Lorenz Hart. Produced by Florenz Ziegfeld & Ed Wynn (uncredited); directed by Zeke Colvan; dances by Seymour Felix.

Cast included: Ruth Etting, Harriet Hoctor, Doree Leslie, Alan Edwards, Lennox Pawle, Bobbe Arnst.
New York run: Ziegfeld Theatre, Feb. 18, 1930; 135 performances.

1931. THE LAUGH PARADE

Variety show with dialogue by Ed Wynn & Ed Preble; music by Harry Warren; lyrics by Mort Dixon & Joe Young. Produced by Ed Wynn; directed by Ed Wynn, Edgar MacGregor; dances by Albertina Rasch.

Cast included: Jeanne Aubert, Eunice Healey, Jack Powell, Lawrence Gray, Bartlett Simmons, The DiGatanos.
New York run: Imperial Theatre, Nov. 2, 1931; 231 performances.

1937. HOORAY FOR WHAT!

Musical comedy with book by Howard Lindsay & Russel Crouse based on idea by E. Y. Harburg; music by Harold Arlen; lyrics by E. Y. Harburg. Produced by the Messrs. Shubert; directed by Vincente Minnelli, Howard Lindsay; dances by Robert Alton, Agnes de Mille.

Cast included: Paul Haakon, June Clyde, Vivian Vance, Jack Whiting, The Briants, Al Gordon's Dogs, Ruthanna Boris, Hugh Martin, Ralph Blane, Meg Mundy.
New York run: Winter Garden, Dec. 1, 1937; 200 performances.

1940. BOYS AND GIRLS TOGETHER

Variety show with dialogue by Ed Wynn & Pat C. Flick; music by Sammy Fain; lyrics by Jack Yellen & Irving Kahal. Produced & directed by Ed Wynn; dances by Albertina Rasch.

Cast included: Renee & Tony DeMarco, Jane Pickens, Dave Apollon, Jerry Cooper, Paul & Frank LaVarre, Dick & Dot Remy.
New York run: Broadhurst Theatre, Oct. 1, 1940; 191 performances.

1942. LAUGH, TOWN, LAUGH

Variety show with dialogue by Ed Wynn (uncredited). Produced & directed by Ed Wynn.

Cast included: Jane Froman, Carmen Amaya, Sabicas, Smith & Dale, Señor Wences, Hermanos Williams Trio, Emil Coleman orch.
New York run: Alvin Theatre, June 22, 1942; 65 performances.

1943. BIG TIME

Variety show with dialogue by Ed Wynn (uncredited). Produced by Fred Finklehoffe & Paul Small; directed by Ed Wynn.

Cast included: Jane Pickens, Paul Draper, Paul & Frank LaVarre, Corinna Mura, Billy Raye.
New York run: None.
NOTE: Show opened at the Mayan Theatre, Los Angeles, April 24, 1943; reopened at the Colonial Theatre, Boston, Aug. 30, 1943, and played two weeks.

1948. ED WYNN'S LAUGH CARNIVAL

Variety show with dialogue by Ed Wynn (uncredited). Produced by Paul Small; directed by Paul Small & Ed Wynn.

Cast included: Phil Baker, Allan Jones, Pat Rooney, Sr., Sid Silvers, Marion Harris, Jr., Hermanos Williams Trio, Dick & Dot Remy.
New York run: None.
NOTE: Show opened at the Curran Theatre, San Francisco, Nov. 7, 1948, and closed at the American Theatre, St. Louis, Jan. 22, 1949.

Index